Undergraduate Texts in Mathematics

Editors

S. Axler
F.W. Gehring
K.A. Ribet

Springer
New York
Berlin
Heidelberg
Barcelona
Budapest
Hong Kong
London
Milan
Paris
Santa Clara
Singapore
Tokyo

Paul R. Halmos

Naive Set Theory

 Springer

P.R. Halmos
Department of Mathematics
Santa Clara University
Santa Clara, CA 95128
USA

Mathematics Subject Classification (1991): 04-01

Halmos, Paul Richard, 1914–
    Naive set theory.

    (Undergraduate texts in mathematics)
    Reprint of the ed. published by Van Nostrand,
Princeton, N.J., in series: The University series
in undergraduate mathematics.
    1. Set theory.   2. Arithmetic—Foundations.
I. Title.
[QA248.H26   1974]   511´.3   74-10687
ISBN 0-387-90092-6

QA
248
H26
1974

Printed on acid-free paper.

Printed and bound by Hamilton Printing Co., Rensselaer, NY.
Printed in the United States of America.

9 8

ISBN 0-387-90092-6 Springer-Verlag New York Berlin Heidelberg
ISBN 3-540-90092-6 Springer-Verlag Berlin Heidelberg New York   SPIN 10662197

PREFACE

Every mathematician agrees that every mathematician must know some set theory; the disagreement begins in trying to decide how much is some. This book contains my answer to that question. The purpose of the book is to tell the beginning student of advanced mathematics the basic set-theoretic facts of life, and to do so with the minimum of philosophical discourse and logical formalism. The point of view throughout is that of a prospective mathematician anxious to study groups, or integrals, or manifolds. From this point of view the concepts and methods of this book are merely some of the standard mathematical tools; the expert specialist will find nothing new here.

Scholarly bibliographical credits and references are out of place in a purely expository book such as this one. The student who gets interested in set theory for its own sake should know, however, that there is much more to the subject than there is in this book. One of the most beautiful sources of set-theoretic wisdom is still Hausdorff's *Set theory*. A recent and highly readable addition to the literature, with an extensive and up-to-date bibliography, is *Axiomatic set theory* by Suppes.

In set theory "naive" and "axiomatic" are contrasting words. The present treatment might best be described as axiomatic set theory from the naive point of view. It is axiomatic in that some axioms for set theory are stated and used as the basis of all subsequent proofs. It is naive in that the language and notation are those of ordinary informal (but formalizable) mathematics. A more important way in which the naive point of view predominates is that set theory is regarded as a body of facts, of which the axioms are a brief and convenient summary; in the orthodox axiomatic view the logical relations among various axioms are the central objects of study. Analogously, a study of geometry might be regarded as purely naive if it proceeded on the paper-folding kind of intuition alone; the other extreme, the purely axiomatic one, is the one in which axioms for the various non-Euclidean geometries are studied with the same amount of attention as Euclid's. The analogue of the point of view of this book is the study of just one sane set of axioms with the intention of describing Euclidean geometry only.

Instead of *Naive set theory* a more honest title for the book would have been *An outline of the elements of naive set theory*. "Elements" would warn the reader that not everything is here; "outline" would warn him that even what is here needs filling in. The style is usually informal to the point of being conversational. There are very few displayed theorems; most of the facts are just stated and followed by a sketch of a proof, very much as they might be in a general descriptive lecture. There are only a few exercises, officially so labelled, but, in fact, most of the book is nothing but a long chain of exercises with hints. The reader should continually ask himself whether he knows how to jump from one hint to the next, and, accordingly, he should not be discouraged if he finds that his reading rate is considerably slower than normal.

This is not to say that the contents of this book are unusually difficult or profound. What is true is that the concepts are very general and very abstract, and that, therefore, they may take some getting used to. It is a mathematical truism, however, that the more generally a theorem applies, the less deep it is. The student's task in learning set theory is to steep himself in unfamiliar but essentially shallow generalities till they become so familiar that they can be used with almost no conscious effort. In other words, general set theory is pretty trivial stuff really, but, if you want to be a mathematician, you need some, and here it is; read it, absorb it, and forget it.

P. R. H.

CONTENTS

SECTION 1

THE AXIOM OF EXTENSION

A pack of wolves, a bunch of grapes, or a flock of pigeons are all examples of sets of things. The mathematical concept of a set can be used as the foundation for all known mathematics. The purpose of this little book is to develop the basic properties of sets. Incidentally, to avoid terminological monotony, we shall sometimes say *collection* instead of *set*. The word "class" is also used in this context, but there is a slight danger in doing so. The reason is that in some approaches to set theory "class" has a special technical meaning. We shall have occasion to refer to this again a little later.

One thing that the development will not include is a definition of sets. The situation is analogous to the familiar axiomatic approach to elementary geometry. That approach does not offer a definition of points and lines; instead it describes what it is that one can do with those objects. The semi-axiomatic point of view adopted here assumes that the reader has the ordinary, human, intuitive (and frequently erroneous) understanding of what sets are; the purpose of the exposition is to delineate some of the many things that one can correctly do with them.

Sets, as they are usually conceived, have *elements* or *members*. An element of a set may be a wolf, a grape, or a pigeon. It is important to know that a set itself may also be an element of some other set. Mathematics is full of examples of sets of sets. A line, for instance, is a set of points; the set of all lines in the plane is a natural example of a set of sets (of points). What may be surprising is not so much that sets may occur as elements, but that for mathematical purposes no other elements need ever be considered. In this book, in particular, we shall study sets, and sets of sets, and similar towers of sometimes frightening height and complexity—and nothing else. By way of examples we might occasionally speak of sets of

1

cabbages, and kings, and the like, but such usage is always to be construed as an illuminating parable only, and not as a part of the theory that is being developed.

The principal concept of set theory, the one that in completely axiomatic studies is the principal primitive (undefined) concept, is that of *belonging*. If x belongs to A (x is an element of A, x is *contained* in A), we shall write

$$x \in A.$$

This version of the Greek letter epsilon is so often used to denote belonging that its use to denote anything else is almost prohibited. Most authors relegate ϵ to its set-theoretic use forever and use ε when they need the fifth letter of the Greek alphabet.

Perhaps a brief digression on alphabetic etiquette in set theory might be helpful. There is no compelling reason for using small and capital letters as in the preceding paragraph; we might have written, and often will write, things like $x \in y$ and $A \in B$. Whenever possible, however, we shall informally indicate the status of a set in a particular hierarchy under consideration by means of the convention that letters at the beginning of the alphabet denote elements, and letters at the end denote sets containing them; similarly letters of a relatively simple kind denote elements, and letters of the larger and gaudier fonts denote sets containing them. Examples: $x \in A$, $A \in X$, $X \in \mathcal{C}$.

A possible relation between sets, more elementary than belonging, is *equality*. The equality of two sets A and B is universally denoted by the familiar symbol

$$A = B;$$

the fact that A and B are not equal is expressed by writing

$$A \neq B.$$

The most basic property of belonging is its relation to equality, which can be formulated as follows.

Axiom of extension. *Two sets are equal if and only if they have the same elements.*

With greater pretentiousness and less clarity: a set is determined by its extension.

It is valuable to understand that the axiom of extension is not just a logically necessary property of equality but a non-trivial statement about belonging. One way to come to understand the point is to consider a partially analogous situation in which the analogue of the axiom of extension does not hold. Suppose, for instance, that we consider human beings instead of sets, and that, if x and A are human beings, we write $x \in A$ whenever x is an ancestor of A. (The ancestors of a human being are his parents, his parents' parents, their parents, etc., etc.) The analogue of the axiom of extension would say here that if two human beings are equal, then they have the same ancestors (this is the "only if" part, and it is true), and also that if two human beings have the same ancestors, then they are equal (this is the "if" part, and it is false).

If A and B are sets and if every element of A is an element of B, we say that A is a *subset* of B, or B *includes* A, and we write

$$A \subset B$$

or

$$B \supset A.$$

The wording of the definition implies that each set must be considered to be included in itself ($A \subset A$); this fact is described by saying that set inclusion is *reflexive*. (Note that, in the same sense of the word, equality also is reflexive.) If A and B are sets such that $A \subset B$ and $A \neq B$, the word *proper* is used (proper subset, proper inclusion). If A, B, and C are sets such that $A \subset B$ and $B \subset C$, then $A \subset C$; this fact is described by saying that set inclusion is *transitive*. (This property is also shared by equality.)

If A and B are sets such that $A \subset B$ and $B \subset A$, then A and B have the same elements and therefore, by the axiom of extension, $A = B$. This fact is described by saying that set inclusion is *antisymmetric*. (In this respect set inclusion behaves differently from equality. Equality is *symmetric*, in the sense that if $A = B$, then necessarily $B = A$.) The axiom of extension can, in fact, be reformulated in these terms: if A and B are sets, then a necessary and sufficient condition that $A = B$ is that both $A \subset B$ and $B \subset A$. Correspondingly, almost all proofs of equalities between two sets A and B are split into two parts; first show that $A \subset B$, and then show that $B \subset A$.

Observe that belonging ($\in$) and inclusion ($\subset$) are conceptually very different things indeed. One important difference has already manifested itself above: inclusion is always reflexive, whereas it is not at all clear that belonging is ever reflexive. That is: $A \subset A$ is always true; is $A \in A$ ever true? It is certainly not true of any reasonable set that anyone has ever seen. Observe, along the same lines, that inclusion is transitive, whereas belonging is not. Everyday examples, involving, for instance, super-organizations whose members are organizations, will readily occur to the interested reader.

SECTION 2

THE AXIOM OF SPECIFICATION

All the basic principles of set theory, except only the axiom of extension, are designed to make new sets out of old ones. The first and most important of these basic principles of set manufacture says, roughly speaking, that anything intelligent one can assert about the elements of a set specifies a subset, namely, the subset of those elements about which the assertion is true.

Before formulating this principle in exact terms, we look at a heuristic example. Let A be the set of all men. The sentence "x is married" is true for some of the elements x of A and false for others. The principle we are illustrating is the one that justifies the passage from the given set A to the subset (namely, the set of all married men) specified by the given sentence. To indicate the generation of the subset, it is usually denoted by

$$\{x \in A : x \text{ is married}\}.$$

Similarly

$$\{x \in A : x \text{ is not married}\}$$

is the set of all bachelors;

$$\{x \in A : \text{the father of } x \text{ is Adam}\}$$

is the set that contains Cain and Abel and nothing else; and

$$\{x \in A : x \text{ is the father of Abel}\}$$

is the set that contains Adam and nothing else. Warning: a box that contains a hat and nothing else is not the same thing as a hat, and, in the same way, the last set in this list of examples is not to be confused with

4

Adam. The analogy between sets and boxes has many weak points, but sometimes it gives a helpful picture of the facts.

All that is lacking for the precise general formulation that underlies the examples above is a definition of *sentence*. Here is a quick and informal one. There are two basic types of sentences, namely, assertions of belonging,

$$x \in A,$$

and assertions of equality,

$$A = B;$$

all other sentences are obtained from such *atomic* sentences by repeated applications of the usual logical operators, subject only to the minimal courtesies of grammar and unambiguity. To make the definition more explicit (and longer) it is necessary to append to it a list of the "usual logical operators" and the rules of syntax. An adequate (and, in fact, redundant) list of the former contains seven items:

and,
or (in the sense of "*either—or—or both*"),
not,
if—then—(or *implies*),
if and only if,
for some (or *there exists*),
for all.

As for the rules of sentence construction, they can be described as follows. (i) Put "not" before a sentence and enclose the result between parentheses. (The reason for parentheses, here and below, is to guarantee unambiguity. Note, incidentally, that they make all other punctuation marks unnecessary. The complete parenthetical equipment that the definition of sentences calls for is rarely needed. We shall always omit as many parentheses as it seems safe to omit without leading to confusion. In normal mathematical practice, to be followed in this book, several different sizes and shapes of parentheses are used, but that is for visual convenience only.) (ii) Put "and" or "or" or "if and only if" between two sentences and enclose the result between parentheses. (iii) Replace the dashes in "if—then —" by sentences and enclose the result in parentheses. (iv) Replace the dash in "for some—" or in "for all—" by a letter, follow the result by a sentence, and enclose the whole in parentheses. (If the letter used does not occur in the sentence, no harm is done. According to the usual and natural convention "for some y ($x \in A$)" just means "$x \in A$". It is equally harmless if the letter used has already been used with "for some—" or "for all—." Recall that "for some x $(x \in A)$" means the same as "for some y $(y \in A)$"; it follows that a judicious change of notation will always avert alphabetic collisions.)

We are now ready to formulate the major principle of set theory, often referred to by its German name *Aussonderungsaxiom*.

Axiom of specification. *To every set A and to every condition $S(x)$ there corresponds a set B whose elements are exactly those elements x of A for which $S(x)$ holds.*

A "condition" here is just a sentence. The symbolism is intended to indicate that the letter x is *free* in the sentence $S(x)$; that means that x occurs in $S(x)$ at least once without being introduced by one of the phrases "for some x" or "for all x." It is an immediate consequence of the axiom of extension that the axiom of specification determines the set B uniquely. To indicate the way B is obtained from A and from $S(x)$ it is customary to write

$$B = \{x \in A : S(x)\}.$$

To obtain an amusing and instructive application of the axiom of specification, consider, in the role of $S(x)$, the sentence

$$\text{not } (x \in x).$$

It will be convenient, here and throughout, to write "$x \in' A$" (alternatively "$x \notin A$") instead of "not $(x \in A)$"; in this notation, the role of $S(x)$ is now played by

$$x \in' x.$$

It follows that, whatever the set A may be, if $B = \{x \in A : x \in' x\}$, then, for all y,

(∗) $y \in B$ if and only if $(y \in A$ and $y \in' y)$.

Can it be that $B \in A$? We proceed to prove that the answer is no. Indeed, if $B \in A$, then either $B \in B$ also (unlikely, but not obviously impossible), or else $B \in' B$. If $B \in B$, then, by (∗), the assumption $B \in A$ yields $B \in' B$—a contradiction. If $B \in' B$, then, by (∗) again, the assumption $B \in A$ yields $B \in B$—a contradiction again. This completes the proof that $B \in A$ is impossible, so that we must have $B \in' A$. The most interesting part of this conclusion is that there exists something (namely B) that does not belong to A. The set A in this argument was quite arbitrary. We have proved, in other words, that

nothing contains everything,

or, more spectacularly,

there is no universe.

"Universe" here is used in the sense of "universe of discourse," meaning, in any particular discussion, a set that contains all the objects that enter into that discussion.

In older (pre-axiomatic) approaches to set theory, the existence of a universe was taken for granted, and the argument in the preceding paragraph was known as the *Russell paradox*. The moral is that it is impossible, especially in mathematics, to get something for nothing. To specify a set, it is not enough to pronounce some magic words (which may form a sentence such as "$x \in' x$"); it is necessary also to have at hand a set to whose elements the magic words apply.

SECTION 3

UNORDERED PAIRS

For all that has been said so far, we might have been operating in a vacuum. To give the discussion some substance, let us now officially assume that

there exists a set.

Since later on we shall formulate a deeper and more useful existential assumption, this assumption plays a temporary role only. One consequence of this innocuous seeming assumption is that there exists a set without any elements at all. Indeed, if A is a set, apply the axiom of specification to A with the sentence "$x \neq x$" (or, for that matter, with any other universally false sentence). The result is the set $\{x \in A : x \neq x\}$, and that set, clearly, has no elements. The axiom of extension implies that there can be only one set with no elements. The usual symbol for that set is

$$\varnothing;$$

the set is called the *empty set*.

The empty set is a subset of every set, or, in other words, $\varnothing \subset A$ for every A. To establish this, we might argue as follows. It is to be proved that every element in $\varnothing$ belongs to A; since there are no elements in $\varnothing$, the condition is automatically fulfilled. The reasoning is correct but perhaps unsatisfying. Since it is a typical example of a frequent phenomenon, a condition holding in the "vacuous" sense, a word of advice to the inexperienced reader might be in order. To prove that something is true about the empty set, prove that it cannot be false. How, for instance, could it be false that $\varnothing \subset A$? It could be false only if $\varnothing$ had an element that did not belong to A. Since $\varnothing$ has no elements at all, this is absurd. Conclusion: $\varnothing \subset A$ is not false, and therefore $\varnothing \subset A$ for every A.

8

The set theory developed so far is still a pretty poor thing; for all we know there is only one set and that one is empty. Are there enough sets to ensure that every set is an element of some set? Is it true that for any two sets there is a third one that they both belong to? What about three sets, or four, or any number? We need a new principle of set construction to resolve such questions. The following principle is a good beginning.

Axiom of pairing. *For any two sets there exists a set that they both belong to.*

Note that this is just the affirmative answer to the second question above.

To reassure worriers, let us hasten to observe that words such as "two," "three," and "four," used above, do not refer to the mathematical concepts bearing those names, which will be defined later; at present such words are merely the ordinary linguistic abbreviations for "something and then something else" repeated an appropriate number of times. Thus, for instance, the axiom of pairing, in unabbreviated form, says that if a and b are sets, then there exists a set A such that $a \in A$ and $b \in A$.

One consequence (in fact an equivalent formulation) of the axiom of pairing is that for any two sets there exists a set that contains both of them and nothing else. Indeed, if a and b are sets, and if A is a set such that $a \in A$ and $b \in A$, then we can apply the axiom of specification to A with the sentence "$x = a$ *or* $x = b$." The result is the set

$$\{x \in A : x = a \text{ or } x = b\},$$

and that set, clearly, contains just a and b. The axiom of extension implies that there can be only one set with this property. The usual symbol for that set is

$$\{a, b\};$$

the set is called the *pair* (or, by way of emphatic comparison with a subsequent concept, the *unordered pair*) formed by a and b.

If, temporarily, we refer to the sentence "$x = a$ *or* $x = b$" as $S(x)$, we may express the axiom of pairing by saying that there exists a set B such that

(*) $x \in B$ *if and only if* $S(x)$.

The axiom of specification, applied to a set A, asserts the existence of a set B such that

(**) $x \in B$ *if and only if* $(x \in A \text{ and } S(x))$.

The relation between (∗) and (∗∗) typifies something that occurs quite frequently. All the remaining principles of set construction are pseudo-special cases of the axiom of specification in the sense in which (∗) is a pseudo-special case of (∗∗). They all assert the existence of a set specified by a certain condition; if it were known in advance that there exists a set containing all the specified elements, then the existence of a set containing just them would indeed follow as a special case of the axiom of specification.

If a is a set, we may form the unordered pair $\{a, a\}$. That unordered pair is denoted by

$$\{a\}$$

and is called the *singleton* of a; it is uniquely characterized by the statement that it has a as its only element. Thus, for instance, $\varnothing$ and $\{\varnothing\}$ are very different sets; the former has no elements, whereas the latter has the unique element $\varnothing$. To say that $a \in A$ is equivalent to saying that $\{a\} \subset A$.

The axiom of pairing ensures that every set is an element of some set and that any two sets are simultaneously elements of some one and the same set. (The corresponding questions for three and four and more sets will be answered later.) Another pertinent comment is that from the assumptions we have made so far we can infer the existence of very many sets indeed. For examples consider the sets $\varnothing$, $\{\varnothing\}$, $\{\{\varnothing\}\}$, $\{\{\{\varnothing\}\}\}$, etc.; consider the pairs, such as $\{\varnothing, \{\varnothing\}\}$, formed by any two of them; consider the pairs formed by any two such pairs, or else the mixed pairs formed by any singleton and any pair; and proceed so on ad infinitum.

EXERCISE. Are all the sets obtained in this way distinct from one another?

Before continuing our study of set theory, we pause for a moment to discuss a notational matter. It seems natural to denote the set B described in (∗) by $\{x : S(x)\}$; in the special case that was there considered

$$\{x : x = a \text{ or } x = b\} = \{a, b\}.$$

We shall use this symbolism whenever it is convenient and permissible to do so. If, that is, $S(x)$ is a condition on x such that the x's that $S(x)$ specifies constitute a set, then we may denote that set by

$$\{x : S(x)\}.$$

In case A is a set and $S(x)$ is $(x \in A)$, then it is permissible to form $\{x : S(x)\}$; in fact

$$\{x : x \in A\} = A.$$

If A is a set and $S(x)$ is an arbitrary sentence, it is permissible to form $\{x: x \in A \text{ and } S(x)\}$; this set is the same as $\{x \in A: S(x)\}$. As further examples, we note that

$$\{x: x \neq x\} = \varnothing$$

and

$$\{x: x = a\} = \{a\}.$$

In case $S(x)$ is $(x \in' x)$, or in case $S(x)$ is $(x = x)$, the specified x's do not constitute a set.

Despite the maxim about never getting something for nothing, it seems a little harsh to be told that certain sets are not really sets and even their names must never be mentioned. Some approaches to set theory try to soften the blow by making systematic use of such illegal sets but just not calling them sets; the customary word is "class." A precise explanation of what classes really are and how they are used is irrelevant in the present approach. Roughly speaking, a class may be identified with a condition (sentence), or, rather, with the "extension" of a condition.

SECTION 4

UNIONS AND INTERSECTIONS

If A and B are sets, it is sometimes natural to wish to unite their elements into one comprehensive set. One way of describing such a comprehensive set is to require it to contain all the elements that belong to at least one of the two members of the pair $\{A, B\}$. This formulation suggests a sweeping generalization of itself; surely a similar construction should apply to arbitrary collections of sets and not just to pairs of them. What is wanted, in other words, is the following principle of set construction.

Axiom of unions. *For every collection of sets there exists a set that contains all the elements that belong to at least one set of the given collection.*

Here it is again: for every collection $\mathcal{C}$ there exists a set U such that if $x \, \epsilon \, X$ for some X in $\mathcal{C}$, then $x \, \epsilon \, U$. (Note that "at least one" is the same as "some.")

The comprehensive set U described above may be too comprehensive; it may contain elements that belong to none of the sets X in the collection $\mathcal{C}$. This is easy to remedy; just apply the axiom of specification to form the set

$$\{x \, \epsilon \, U : x \, \epsilon \, X \text{ for some } X \text{ in } \mathcal{C}\}.$$

(The condition here is a translation into idiomatic usage of the mathematically more acceptable *"for some X ($x \, \epsilon \, X$ and $X \, \epsilon \, \mathcal{C}$).")* It follows that, for every x, a necessary and sufficient condition that x belong to this set is that x belong to X for some X in $\mathcal{C}$. If we change notation and call the new set U again, then

$$U = \{x : x \, \epsilon \, X \text{ for some } X \text{ in } \mathcal{C}\}.$$

This set U is called the *union* of the collection $\mathcal{C}$ of sets; note that the

12

axiom of extension guarantees its uniqueness. The simplest symbol for U that is in use at all is not very popular in mathematical circles; it is

$$\bigcup \mathfrak{C}.$$

Most mathematicians prefer something like

$$\bigcup \{X : X \in \mathfrak{C}\}$$

or

$$\bigcup_{X \in \mathfrak{C}} X.$$

Further alternatives are available in certain important special cases; they will be described in due course.

For the time being we restrict our study of the theory of unions to the simplest facts only. The simplest fact of all is that

$$\bigcup \{X : X \in \varnothing\} = \varnothing,$$

and the next simplest fact is that

$$\bigcup \{X : X \in \{A\}\} = A.$$

In the brutally simple notation mentioned above these facts are expressed by

$$\bigcup \varnothing = \varnothing$$

and

$$\bigcup \{A\} = A.$$

The proofs are immediate from the definitions.

There is a little more substance in the union of pairs of sets (which is what started this whole discussion anyway). In that case special notation is used:

$$\bigcup \{X : X \in \{A, B\}\} = A \cup B.$$

The general definition of unions implies in the special case that $x \in A \cup B$ if and only if x belongs to either A or B or both; it follows that

$$A \cup B = \{x : x \in A \text{ or } x \in B\}.$$

Here are some easily proved facts about the unions of pairs:

$$A \cup \varnothing = A,$$

$$A \cup B = B \cup A \text{ (commutativity)},$$

$$A \cup (B \cup C) = (A \cup B) \cup C \text{ (associativity)},$$

$$A \cup A = A \text{ (idempotence)},$$

$$A \subset B \text{ if and only if } A \cup B = B.$$

Every student of mathematics should prove these things for himself at least once in his life. The proofs are based on the corresponding elementary properties of the logical operator *or*.

An equally simple but quite suggestive fact is that

$$\{a\} \cup \{b\} = \{a, b\}.$$

What this suggests is the way to generalize pairs. Specifically, we write

$$\{a, b, c\} = \{a\} \cup \{b\} \cup \{c\}.$$

The equation defines its left side. The right side should by rights have at least one pair of parentheses in it, but, in view of the associative law, their omission can lead to no misunderstanding. Since it is easy to prove that

$$\{a, b, c\} = \{x: x = a \text{ or } x = b \text{ or } x = c\},$$

we know now that for every three sets there exists a set that contains them and nothing else; it is natural to call that uniquely determined set the (*unordered*) *triple* formed by them. The extension of the notation and terminology thus introduced to more terms (*quadruples*, etc.) is obvious.

The formation of unions has many points of similarity with another set-theoretic operation. If A and B are sets, the *intersection* of A and B is the set

$$A \cap B$$

defined by

$$A \cap B = \{x \in A : x \in B\}.$$

The definition is symmetric in A and B even if it looks otherwise; we have

$$A \cap B = \{x \in B : x \in A\},$$

and, in fact, since $x \in A \cap B$ if and only if x belongs to both A and B, it follows that

$$A \cap B = \{x: x \in A \text{ and } x \in B\}.$$

The basic facts about intersections, as well as their proofs, are similar to the basic facts about unions:

$$A \cap \varnothing = \varnothing,$$
$$A \cap B = B \cap A,$$
$$A \cap (B \cap C) = (A \cap B) \cap C,$$
$$A \cap A = A,$$
$$A \subset B \text{ if and only if } A \cap B = A.$$

Pairs of sets with an empty intersection occur frequently enough to justify the use of a special word: if $A \cap B = \varnothing$, the sets A and B are called *disjoint*. The same word is sometimes applied to a collection of sets to indicate that any two distinct sets of the collection are disjoint; alternatively we may speak in such a situation of a *pairwise disjoint* collection.

Two useful facts about unions and intersections involve both the operations at the same time:

$$A \cap (B \cup C) = (A \cap B) \cup (A \cap C),$$

$$A \cup (B \cap C) = (A \cup B) \cap (A \cup C).$$

These identities are called the *distributive laws*. By way of a sample of a set-theoretic proof, we prove the second one. If x belongs to the left side, then x belongs either to A or to both B and C; if x is in A, then x is in both $A \cup B$ and $A \cup C$, and if x is in both B and C, then, again, x is in both $A \cup B$ and $A \cup C$; it follows that, in any case, x belongs to the right side. This proves that the right side includes the left. To prove the reverse inclusion, just observe that if x belongs to both $A \cup B$ and $A \cup C$, then x belongs either to A or to both B and C.

The formation of the intersection of two sets A and B, or, we might as well say, the formation of the intersection of a pair $\{A, B\}$ of sets, is a special case of a much more general operation. (This is another respect in which the theory of intersections imitates that of unions.) The existence of the general operation of intersection depends on the fact that for each non-empty collection of sets there exists a set that contains exactly those elements that belong to every set of the given collection. In other words: for each collection $\mathcal{C}$, other than $\varnothing$, there exists a set V such that $x \in V$ if and only if $x \in X$ for every X in $\mathcal{C}$. To prove this assertion, let A be any particular set in $\mathcal{C}$ (this step is justified by the fact that $\mathcal{C} \neq \varnothing$) and write

$$V = \{x \in A : x \in X \text{ for every } X \text{ in } \mathcal{C}\}.$$

(The condition means "*for all* X (*if* $X \in \mathcal{C}$, *then* $x \in X$).") The dependence of V on the arbitrary choice of A is illusory; in fact

$$V = \{x : x \in X \text{ for every } X \text{ in } \mathcal{C}\}.$$

The set V is called the *intersection* of the collection $\mathcal{C}$ of sets; the axiom of extension guarantees its uniqueness. The customary notation is similar to the one for unions: instead of the unobjectionable but unpopular

$$\bigcap \mathcal{C},$$

the set V is usually denoted by

$$\bigcap \{X : X \in \mathfrak{C}\}$$

or

$$\bigcap_{x \in \mathfrak{C}} X.$$

EXERCISE. A necessary and sufficient condition that $(A \cap B) \cup C = A \cap (B \cup C)$ is that $C \subset A$. Observe that the condition has nothing to do with the set B.

SECTION 5

COMPLEMENTS AND POWERS

If A and B are sets, the *difference* between A and B, more often known as the *relative complement* of B in A, is the set $A - B$ defined by

$$A - B = \{x \,\epsilon\, A : x \,\epsilon'\, B\}.$$

Note that in this definition it is not necessary to assume that $B \subset A$. In order to record the basic facts about complementation as simply as possible, we assume nevertheless (in this section only) that all the sets to be mentioned are subsets of one and the same set E and that all complements (unless otherwise specified) are formed relative to that E. In such situations (and they are quite common) it is easier to remember the underlying set E than to keep writing it down, and this makes it possible to simplify the notation. An often used symbol for the temporarily absolute (as opposed to relative) complement of A is A'. In terms of this symbol the basic facts about complementation can be stated as follows:

$$(A')' = A,$$

$$\varnothing' = E, \quad E' = \varnothing,$$

$$A \cap A' = \varnothing, \quad A \cup A' = E,$$

$$A \subset B \text{ if and only if } B' \subset A'.$$

The most important statements about complements are the so-called *De Morgan laws:*

$$(A \cup B)' = A' \cap B', \quad (A \cap B)' = A' \cup B'.$$

(We shall see presently that the De Morgan laws hold for the unions and intersections of larger collections of sets than just pairs.) These facts about

17

complementation imply that the theorems of set theory usually come in pairs. If in an inclusion or equation involving unions, intersections, and complements of subsets of E we replace each set by its complement, interchange unions and intersections, and reverse all inclusions, the result is another theorem. This fact is sometimes referred to as the *principle of duality* for sets.

Here are some easy exercises on complementation.

$$A - B = A \cap B'.$$

$$A \subset B \text{ if and only if } A - B = \varnothing.$$

$$A - (A - B) = A \cap B.$$

$$A \cap (B - C) = (A \cap B) - (A \cap C).$$

$$A \cap B \subset (A \cap C) \cup (B \cap C').$$

$$(A \cup C) \cap (B \cup C') \subset A \cup B.$$

If A and B are sets, the *symmetric difference* (or *Boolean sum*) of A and B is the set $A + B$ defined by

$$A + B = (A - B) \cup (B - A).$$

This operation is commutative ($A + B = B + A$) and associative ($A + (B + C) = (A + B) + C$), and is such that $A + \varnothing = A$ and $A + A = \varnothing$.

This may be the right time to straighten out a trivial but occasionally puzzling part of the theory of intersections. Recall, to begin with, that intersections were defined for non-empty collections only. The reason is that the same approach to the empty collection does not define a set. Which x's are specified by the sentence

$$x \in X \text{ for every } X \text{ in } \varnothing?$$

As usual for questions about $\varnothing$ the answer is easier to see for the corresponding negative question. Which x's do *not* satisfy the stated condition? If it is not true that $x \in X$ for every X in $\varnothing$, then there must exist an X in $\varnothing$ such that $x \in' X$; since, however, there do not exist any X's in $\varnothing$ at all, this is absurd. Conclusion: no x fails to satisfy the stated condition, or, equivalently, every x does satisfy it. In other words, the x's that the condition specifies exhaust the (nonexistent) universe. There is no profound problem here; it is merely a nuisance to be forced always to be making qualifications and exceptions just because some set somewhere along some construction might turn out to be empty. There is nothing to be done about this; it is just a fact of life.

If we restrict our attention to subsets of a particular set E, as we have temporarily agreed to do, then the unpleasantness described in the preceding paragraph appears to go away. The point is that in that case we can define the intersection of a collection $\mathcal{C}$ (of subsets of E) to be the set

$$\{x \in E: x \in X \text{ for every } X \text{ in } \mathcal{C}\}.$$

This is nothing revolutionary; for each non-empty collection, the new definition agrees with the old one. The difference is in the way the old and the new definitions treat the empty collection; according to the new definition $\bigcap_{x \in \varnothing} X$ is equal to E. (For which elements x of E can it be false that $x \in X$ for every X in $\varnothing$?) The difference is just a matter of language. A little reflection reveals that the "new" definition offered for the intersection of a collection $\mathcal{C}$ of subsets of E is really the same as the old definition of the intersection of the collection $\mathcal{C} \cup \{E\}$, and the latter is never empty.

We have been considering the subsets of a set E; do those subsets themselves constitute a set? The following principle guarantees that the answer is yes.

Axiom of powers. *For each set there exists a collection of sets that contains among its elements all the subsets of the given set.*

In other words, if E is a set, then there exists a set (collection) $\mathcal{P}$ such that if $X \subset E$, then $X \in \mathcal{P}$.

The set $\mathcal{P}$ described above may be larger than wanted; it may contain elements other than the subsets of E. This is easy to remedy; just apply the axiom of specification to form the set $\{X \in \mathcal{P}: X \subset E\}$. (Recall that "$X \subset E$" says the same thing as *"for all x (if $x \in X$ then $x \in E$)."*) Since, for every X, a necessary and sufficient condition that X belong to this set is that X be a subset of E, it follows that if we change notation and call this set $\mathcal{P}$ again, then

$$\mathcal{P} = \{X: X \subset E\}.$$

The set $\mathcal{P}$ is called the *power set* of E; the axiom of extension guarantees its uniqueness. The dependence of $\mathcal{P}$ on E is denoted by writing $\mathcal{P}(E)$ instead of just $\mathcal{P}$.

Because the set $\mathcal{P}(E)$ is very big in comparison with E, it is not easy to give examples. If $E = \varnothing$, the situation is clear enough; the set $\mathcal{P}(\varnothing)$ is the singleton $\{\varnothing\}$. The power sets of singletons and pairs are also easily describable; we have

$$\mathcal{P}(\{a\}) = \{\varnothing, \{a\}\}$$

and

$$\mathcal{P}(\{a, b\}) = \{\varnothing, \{a\}, \{b\}, \{a, b\}\}.$$

The power set of a triple has eight elements. The reader can probably guess (and is hereby challenged to prove) the generalization that includes all these statements: the power set of a finite set with, say, n elements has 2^n elements. (Of course concepts like "finite" and "2^n" have no official standing for us yet; this should not prevent them from being unofficially understood.) The occurrence of n as an exponent (the n-th power of 2) has something to do with the reason why a power set bears its name.

If $\mathcal{C}$ is a collection of subsets of a set E (that is, $\mathcal{C}$ is a subcollection of $\mathcal{P}(E)$), then write

$$\mathcal{D} = \{X \in \mathcal{P}(E) : X' \in \mathcal{C}\}.$$

(To be certain that the condition used in the definition of $\mathcal{D}$ is a sentence in the precise technical sense, it must be rewritten in something like the form

for some Y [$Y \in \mathcal{C}$ and for all x ($x \in X$ if and only if ($x \in E$ and $x \in' Y$))].

Similar comments often apply when we wish to use defined abbreviations instead of logical and set-theoretic primitives only. The translation rarely requires any ingenuity and we shall usually omit it.) It is customary to denote the union and the intersection of the collection $\mathcal{D}$ by the symbols

$$\bigcup_{x \in \mathcal{C}} X' \quad \text{and} \quad \bigcap_{x \in \mathcal{C}} X'.$$

In this notation the general forms of the De Morgan laws become

$$\left(\bigcup_{x \in \mathcal{C}} X\right)' = \bigcap_{x \in \mathcal{C}} X'$$

and

$$\left(\bigcap_{x \in \mathcal{C}} X\right)' = \bigcup_{x \in \mathcal{C}} X'.$$

The proofs of these equations are immediate consequences of the appropriate definitions.

EXERCISE. Prove that $\mathcal{P}(E) \cap \mathcal{P}(F) = \mathcal{P}(E \cap F)$ and $\mathcal{P}(E) \cup \mathcal{P}(F) \subset \mathcal{P}(E \cup F)$. These assertions can be generalized to

$$\bigcap_{x \in \mathcal{C}} \mathcal{P}(X) = \mathcal{P}\left(\bigcap_{x \in \mathcal{C}} X\right)$$

and

$$\bigcup_{x \in \mathcal{C}} \mathcal{P}(X) \subset \mathcal{P}\left(\bigcup_{x \in \mathcal{C}} X\right);$$

find a reasonable interpretation of the notation in which these generalizations were here expressed and then prove them. Further elementary
facts:

$$\bigcap_{X \,\in\, \mathcal{P}(E)} X = \varnothing,$$

and

$$\text{if } E \subset F, \text{ then } \mathcal{P}(E) \subset \mathcal{P}(F).$$

A curious question concerns the commutativity of the operators $\mathcal{P}$ and
$\bigcup$. Show that E is always equal to $\bigcup_{X \,\in\, \mathcal{P}(E)} X$ (that is $E = \bigcup \mathcal{P}(E)$),
but that the result of applying $\mathcal{P}$ and $\bigcup$ to E in the other order is a set
that includes E as a subset, typically a proper subset.

SECTION 6

ORDERED PAIRS

What does it mean to arrange the elements of a set A in some order? Suppose, for instance, that the set A is the quadruple $\{a, b, c, d\}$ of distinct elements, and suppose that we want to consider its elements in the order

$$c \; b \; d \; a.$$

Even without a precise definition of what this means, we can do something set-theoretically intelligent with it. We can, namely, consider, for each particular spot in the ordering, the set of all those elements that occur at or before that spot; we obtain in this way the sets

$$\{c\} \quad \{c, b\} \quad \{c, b, d\} \quad \{c, b, d, a\}.$$

We can go on then to consider the set (or collection, if that sounds better)

$$\mathcal{C} = \{\{a, b, c, d\}, \{b, c\}, \{b, c, d\}, \{c\}\}$$

that has exactly those sets for its elements. In order to emphasize that the intuitively based and possibly unclear concept of order has succeeded in producing something solid and simple, namely a plain, unembellished set $\mathcal{C}$, the elements of $\mathcal{C}$, and *their* elements, are presented above in a scrambled manner. (The lexicographically inclined reader might be able to see a method in the manner of scrambling.)

Let us continue to pretend for a while that we do know what order means. Suppose that in a hasty glance at the preceding paragraph all we could catch is the set $\mathcal{C}$; can we use it to recapture the order that gave rise to it? The answer is easily seen to be yes. Examine the elements of $\mathcal{C}$ (they themselves are sets, of course) to find one that is included in all the others; since $\{c\}$ fills the bill (and nothing else does) we know that c must have been the first element. Look next for the next smallest element of $\mathcal{C}$, i.e., the one that is included in all the ones that remain after $\{c\}$ is removed; since $\{b, c\}$ fills the bill (and nothing else does), we know that b must have been the second element. Proceeding thus (only two more steps are needed) we pass from the set $\mathcal{C}$ to the given ordering of the given set A.

The moral is this: we may not know precisely what it means to order the elements of a set A, but with each order we can associate a set $\mathcal{C}$ of subsets of A in such a way that the given order can be uniquely recaptured from $\mathcal{C}$. (Here is a non-trivial exercise: find an intrinsic characterization of those sets of subsets of A that correspond to some order in A. Since "order" has no official meaning for us yet, the whole problem is officially meaningless. Nothing that follows depends on the solution, but the reader would learn something valuable by trying to find it.) The passage from an order in A to the set $\mathcal{C}$, and back, was illustrated above for a quadruple; for a pair everything becomes at least twice as simple. If $A = \{a, b\}$ and if, in the desired order, a comes first, then $\mathcal{C} = \{\{a\}, \{a, b\}\}$; if, however, b comes first, then $\mathcal{C} = \{\{b\}, \{a, b\}\}$.

The *ordered pair* of a and b, with *first coordinate* a and *second coordinate* b, is the set (a, b) defined by

$$(a, b) = \{\{a\}, \{a, b\}\}.$$

However convincing the motivation of this definition may be, we must still prove that the result has the main property that an ordered pair must have to deserve its name. We must show that if (a, b) and (x, y) are ordered pairs and if $(a, b) = (x, y)$, then $a = x$ and $b = y$. To prove this, we note first that if a and b happen to be equal, then the ordered pair (a, b) is the same as the singleton $\{\{a\}\}$. If, conversely, (a, b) is a singleton, then $\{a\} = \{a, b\}$, so that $b \in \{a\}$, and therefore $a = b$. Suppose now that $(a, b) = (x, y)$. If $a = b$, then both (a, b) and (x, y) are singletons, so that $x = y$; since $\{x\} \in (a, b)$ and $\{a\} \in (x, y)$, it follows that a, b, x, and y are all equal. If $a \neq b$, then both (a, b) and (x, y) contain exactly one singleton, namely $\{a\}$ and $\{x\}$ respectively, so that $a = x$. Since in this case it is also true that both (a, b) and (x, y) contain exactly one unordered pair that is not a singleton, namely $\{a, b\}$ and $\{x, y\}$ respectively, it follows that $\{a, b\} = \{x, y\}$, and therefore, in particular, $b \in \{x, y\}$. Since b cannot be x (for then we should have $a = x$ and $b = x$, and, therefore, $a = b$), we must have $b = y$, and the proof is complete.

If A and B are sets, does there exist a set that contains all the ordered pairs (a, b) with a in A and b in B? It is quite easy to see that the answer is yes. Indeed, if $a \in A$ and $b \in B$, then $\{a\} \subset A$ and $\{b\} \subset B$, and therefore $\{a, b\} \subset A \cup B$. Since also $\{a\} \subset A \cup B$, it follows that both $\{a\}$

and $\{a, b\}$ are elements of $\mathcal{O}(A \cup B)$. This implies that $\{\{a\}, \{a, b\}\}$ is a subset of $\mathcal{O}(A \cup B)$, and hence that it is an element of $\mathcal{O}(\mathcal{O}(A \cup B))$; in other words $(a, b) \in \mathcal{O}(\mathcal{O}(A \cup B))$ whenever $a \in A$ and $b \in B$. Once this is known, it is a routine matter to apply the axiom of specification and the axiom of extension to produce the unique set $A \times B$ that consists exactly of the ordered pairs (a, b) with a in A and b in B. This set is called the *Cartesian product* of A and B; it is characterized by the fact that

$$A \times B = \{x: x = (a, b) \text{ for some } a \text{ in } A \text{ and for some } b \text{ in } B\}.$$

The Cartesian product of two sets is a set of ordered pairs (that is, a set each of whose elements is an ordered pair), and the same is true of every subset of a Cartesian product. It is of technical importance to know that we can go in the converse direction also: every set of ordered pairs is a subset of the Cartesian product of two sets. In other words: if R is a set such that every element of R is an ordered pair, then there exist two sets A and B such that $R \subset A \times B$. The proof is elementary. Suppose indeed that $x \in R$, so that $x = \{\{a\}, \{a, b\}\}$ for some a and for some b. The problem is to dig out a and b from under the braces. Since the elements of R are sets, we can form the union of the sets in R; since x is one of the sets in R, the elements of x belong to that union. Since $\{a, b\}$ is one of the elements of x, we may write, in what has been called the brutal notation above, $\{a, b\} \in \bigcup R$. One set of braces has disappeared; let us do the same thing again to make the other set go away. Form the union of the sets in $\bigcup R$. Since $\{a, b\}$ is one of those sets, it follows that the elements of $\{a, b\}$ belong to that union, and hence both a and b belong to $\bigcup \bigcup R$. This fulfills the promise made above; to exhibit R as a subset of some $A \times B$, we may take both A and B to be $\bigcup \bigcup R$. It is often desirable to take A and B as small as possible. To do so, just apply the axiom of specification to produce the sets

$$A = \{a: \text{for some } b \ ((a, b) \in R)\}$$

and

$$B = \{b: \text{for some } a \ ((a, b) \in R)\}.$$

These sets are called the *projections* of R onto the first and second coordinates respectively.

However important set theory may be now, when it began some scholars considered it a disease from which, it was to be hoped, mathematics would soon recover. For this reason many set-theoretic considerations were called pathological, and the word lives on in mathematical usage; it often refers to something the speaker does not like. The explicit definition of an ordered pair $((a, b) = \{\{a\}, \{a, b\}\})$ is frequently relegated to pathological set theory. For the benefit of those who think that in this case the name is deserved, we note that the definition has served its purpose by now and will never be used again. We need to know that ordered pairs are determined by and uniquely determine their first and second coordinates, that Cartesian products can be formed, and that every set of ordered pairs is a subset of some Cartesian product; which particular approach is used to achieve these ends is immaterial.

It is easy to locate the source of the mistrust and suspicion that many mathematicians feel toward the explicit definition of ordered pair given above. The trouble is not that there is anything wrong or anything missing; the relevant properties of the concept we have defined are all correct (that is, in accord with the demands of intuition) and all the correct properties are present. The trouble is that the concept has some irrelevant properties that are accidental and distracting. The theorem that $(a, b) = (x, y)$ if and only if $a = x$ and $b = y$ is the sort of thing we expect to learn about ordered pairs. The fact that $\{a, b\} \in (a, b)$, on the other hand, seems accidental; it is a freak property of the definition rather than an intrinsic property of the concept.

The charge of artificiality is true, but it is not too high a price to pay for conceptual economy. The concept of an ordered pair could have been introduced as an additional primitive, axiomatically endowed with just the right properties, no more and no less. In some theories this is done. The mathematician's choice is between having to remember a few more axioms and having to forget a few accidental facts; the choice is pretty clearly a matter of taste. Similar choices occur frequently in mathematics; in this book, for instance, we shall encounter them again in connection with the definitions of numbers of various kinds.

EXERCISE. If A, B, X, and Y are sets, then

    (i) $(A \cup B) \times X = (A \times X) \cup (B \times X)$,

    (ii) $(A \cap B) \times (X \cap Y) = (A \times X) \cap (B \times Y)$,

    (iii) $(A - B) \times X = (A \times X) - (B \times X)$.

If either $A = \varnothing$ or $B = \varnothing$, then $A \times B = \varnothing$, and conversely. If $A \subset X$ and $B \subset Y$, then $A \times B \subset X \times Y$, and (provided $A \times B \neq \varnothing$) conversely.

SECTION 7

RELATIONS

Using ordered pairs, we can formulate the mathematical theory of relations in set-theoretic language. By a relation we mean here something like marriage (between men and women) or belonging (between elements and sets). More explicitly, what we shall call a relation is sometimes called a *binary* relation. An example of a ternary relation is parenthood for people (Adam and Eve are the parents of Cain). In this book we shall have no occasion to treat the theory of relations that are ternary, quaternary, or worse.

Looking at any specific relation, such as marriage for instance, we might be tempted to consider certain ordered pairs (x, y), namely just those for which x is a man, y is a woman, and x is married to y. We have not yet seen the definition of the general concept of a relation, but it seems plausible that, just as in this marriage example, every relation should uniquely determine the set of all those ordered pairs for which the first coordinate does stand in that relation to the second. If we know the relation, we know the set, and, better yet, if we know the set, we know the relation. If, for instance, we were presented with the set of ordered pairs of people that corresponds to marriage, then, even if we forgot the definition of marriage, we could always tell when a man x is married to a woman y and when not; we would just have to see whether the ordered pair (x, y) does or does not belong to the set.

We may not know what a relation is, but we do know what a set is, and the preceding considerations establish a close connection between relations and sets. The precise set-theoretic treatment of relations takes advantage of that heuristic connection; the simplest thing to do is to define a relation to be the corresponding set. This is what we do; we hereby define a *relation* as a set of ordered pairs. Explicitly: a set R is a relation if each ele-

26

ment of R is an ordered pair; this means, of course, that if $z \in R$, then there exist x and y so that $z = (x, y)$. If R is a relation, it is sometimes convenient to express the fact that $(x, y) \in R$ by writing

$$x \, R \, y$$

and saying, as in everyday language, that x stands in the relation R to y.

The least exciting relation is the empty one. (To prove that $\varnothing$ is a set of ordered pairs, look for an element of $\varnothing$ that is not an ordered pair.) Another dull example is the Cartesian product of any two sets X and Y. Here is a slightly more interesting example: let X be any set, and let R be the set of all those pairs (x, y) in $X \times X$ for which $x = y$. The relation R is just the relation of equality between elements of X; if x and y are in X, then $x \, R \, y$ means the same as $x = y$. One more example will suffice for now: let X be any set, and let R be the set of all those pairs (x, A) in $X \times \mathcal{O}(X)$ for which $x \in A$. This relation R is just the relation of belonging between elements of X and subsets of X; if $x \in X$ and $A \in \mathcal{O}(X)$, then $x \, R \, A$ means the same as $x \in A$.

In the preceding section we saw that associated with every set R of ordered pairs there are two sets called the projections of R onto the first and second coordinates. In the theory of relations these sets are known as the *domain* and the *range* of R (abbreviated dom R and ran R); we recall that they are defined by

$$\text{dom } R = \{x \colon \textit{for some } y \; (x \, R \, y)\}$$

and

$$\text{ran } R = \{y \colon \textit{for some } x \; (x \, R \, y)\}.$$

If R is the relation of marriage, so that $x \, R \, y$ means that x is a man, y is a woman, and x and y are married to one another, then dom R is the set of married men and ran R is the set of married women. Both the domain and the range of $\varnothing$ are equal to $\varnothing$. If $R = X \times Y$, then dom $R = X$ and ran $R = Y$. If R is equality in X, then dom $R = \text{ran } R = X$. If R is belonging, between X and $\mathcal{O}(X)$, then dom $R = X$ and ran $R = \mathcal{O}(X) - \{\varnothing\}$.

If R is a relation included in a Cartesian product $X \times Y$ (so that dom $R \subset X$ and ran $R \subset Y$), it is sometimes convenient to say that R is a relation *from* X *to* Y; instead of a relation from X to X we may speak of a relation *in* X. A relation R in X is *reflexive* if $x \, R \, x$ for every x in X; it is *symmetric* if $x \, R \, y$ implies that $y \, R \, x$; and it is *transitive* if $x \, R \, y$ and $y \, R \, z$ imply that $x \, R \, z$. (Exercise: for each of these three possible properties, find a relation that does not have that property but does have the other two.) A relation in a set is an *equivalence relation* if it is reflexive, symmetric, and transitive. The smallest equivalence relation in a set X is the relation of equality in X; the largest equivalence relation in X is $X \times X$.

There is an intimate connection between equivalence relations in a set X and certain collections (called partitions) of subsets of X. A *partition* of X is a disjoint collection $\mathcal{C}$ of non-empty subsets of X whose union is X. If R is an equivalence relation in X, and if x is in X, the *equivalence class* of x with respect to R is the set of all those elements y in X for which $x \, R \, y$. (The weight of tradition makes the use of the word "class" at this point unavoidable.) Examples: if R is equality in X, then each equivalence class is a singleton; if $R = X \times X$, then the set X itself is the only equivalence class. There is no standard notation for the equivalence class of x with respect to R; we shall usually denote it by x/R, and we shall write X/R for the set of all equivalence classes. (Pronounce X/R as "X modulo R," or, in abbreviated form, "X mod R." Exercise: show that X/R is indeed a set by exhibiting a condition that specifies exactly the subset X/R of the power set $\mathcal{P}(X)$.) Now forget R for a moment and begin anew with a partition $\mathcal{C}$ of X. A relation, which we shall call $X/\mathcal{C}$, is defined in X by writing

$$x \quad X/\mathcal{C} \quad y$$

just in case x and y belong to the same set of the collection $\mathcal{C}$. We shall call $X/\mathcal{C}$ the relation *induced* by the partition $\mathcal{C}$.

In the preceding paragraph we saw how to associate a set of subsets of X with every equivalence relation in X and how to associate a relation in X with every partition of X. The connection between equivalence relations and partitions can be described by saying that the passage from $\mathcal{C}$ to $X/\mathcal{C}$ is exactly the reverse of the passage from R to X/R. More explicitly: if R is an equivalence relation in X, then the set of equivalence classes is a partition of X that induces the relation R, and if $\mathcal{C}$ is a partition of X, then the induced relation is an equivalence relation whose set of equivalence classes is exactly $\mathcal{C}$.

For the proof, let us start with an equivalence relation R. Since each x belongs to some equivalence class (for instance $x \, \epsilon \, x/R$), it is clear that the union of the equivalence classes is all X. If $z \, \epsilon \, x/R \cap y/R$, then $x \, R \, z$ and $z \, R \, y$, and therefore $x \, R \, y$. This implies that if two equivalence classes have an element in common, then they are identical, or, in other words, that two distinct equivalence classes are always disjoint. The set of equivalence classes is therefore a partition. To say that two elements belong to the same set (equivalence class) of this partition means, by definition, that they stand in the relation R to one another. This proves the first half of our assertion.

The second half is easier. Start with a partition C and consider the induced relation. Since every element of X belongs to some set of C, reflexivity just says that x and x are in the same set of C. Symmetry says that if x and y are in the same set of C, then y and x are in the same set of C, and this is obviously true. Transitivity says that if x and y are in the same set of C and if y and z are in the same set of C, then x and z are in the same set of C, and this too is obvious. The equivalence class of each x in X is just the set of C to which x belongs. This completes the proof of everything that was promised.

SECTION 8

FUNCTIONS

If X and Y are sets, a *function* from (or *on*) X *to* (or *into*) Y is a relation f such that dom $f = X$ and such that for each x in X there is a unique element y in Y with $(x, y) \in f$. The uniqueness condition can be formulated explicitly as follows: if $(x, y) \in f$ and $(x, z) \in f$, then $y = z$. For each x in X, the unique y in Y such that $(x, y) \in f$ is denoted by $f(x)$. For functions this notation and its minor variants supersede the others used for more general relations; from now on, if f is a function, we shall write $f(x) = y$ instead of $(x, y) \in f$ or $x f y$. The element y is called the *value* that the function f *assumes* (or *takes on*) at the *argument* x; equivalently we may say that f *sends* or *maps* or *transforms* x onto y. The words *map* or *mapping*, *transformation*, *correspondence*, and *operator* are among some of the many that are sometimes used as synonyms for *function*. The symbol

$$f\colon X \to Y$$

is sometimes used as an abbreviation for "f is a function from X to Y." The set of all functions from X to Y is a subset of the power set $\mathcal{P}(X \times Y)$; it will be denoted by Y^X.

The connotations of activity suggested by the synonyms listed above make some scholars dissatisfied with the definition according to which a function does not *do* anything but merely *is*. This dissatisfaction is reflected in a different use of the vocabulary: *function* is reserved for the undefined object that is somehow active, and the set of ordered pairs that we have called the function is then called the *graph* of the function. It is easy to find examples of functions in the precise set-theoretic sense of the word in both mathematics and everyday life; all we have to look for is information, not necessarily numerical, in tabulated form. One example

30

is a city directory; the arguments of the function are, in this case, the inhabitants of the city, and the values are their addresses.

For relations in general, and hence for functions in particular, we have defined the concepts of domain and range. The domain of a function f from X into Y is, by definition, equal to X, but its range need not be equal to Y; the range consists of those elements y of Y for which there exists an x in X such that $f(x) = y$. If the range of f is equal to Y, we say that f maps X *onto* Y. If A is a subset of X, we may want to consider the set of all those elements y of Y for which there exists an x in the subset A such that $f(x) = y$. This subset of Y is called the *image* of A under f and is frequently denoted by $f(A)$. The notation is bad but not catastrophic. What is bad about it is that if A happens to be both an element of X and a subset of X (an unlikely situation, but far from an impossible one), then the symbol $f(A)$ is ambiguous. Does it mean the value of f at A or does it mean the set of values of f at the elements of A? Following normal mathematical custom, we shall use the bad notation, relying on context, and, on the rare occasions when it is necessary, adding verbal stipulations, to avoid confusion. Note that the image of X itself is the range of f; the "onto" character of f can be expressed by writing $f(X) = Y$.

If X is a subset of a set Y, the function f defined by $f(x) = x$ for each x in X is called the *inclusion map* (or the *embedding*, or the *injection*) of X into Y. The phrase "the function f defined by . . ." is a very common one in such contexts. It is intended to imply, of course, that there does indeed exist a unique function satisfying the stated condition. In the special case at hand this is obvious enough; we are being invited to consider the set of all those ordered pairs (x, y) in $X \times Y$ for which $x = y$. Similar considerations apply in every case, and, following normal mathematical practice, we shall usually describe a function by describing its value y at each argument x. Such a description is sometimes longer and more cumbersome than a direct description of the set (of ordered pairs) involved, but, nevertheless, most mathematicians regard the argument-value description as more perspicuous than any other.

The inclusion map of X into X is called the *identity map* on X. (In the language of relations, the identity map on X is the same as the relation of equality in X.) If, as before, $X \subset Y$, then there is a connection between the inclusion map of X into Y and the identity map on Y; that connection is a special case of a general procedure for making small functions out of large ones. If f is a function from Y to Z, say, and if X is a subset of Y, then there is a natural way of constructing a function g from X to Z; define $g(x)$ to be equal to $f(x)$ for each x in X. The function g is called the *restriction* of f to X, and f is called an *extension* of g to Y; it is customary to write $g = f \mid X$. The definition of restriction can be expressed by writing $(f \mid X)(x) = f(x)$ for each x in X; observe also that ran $(f \mid X) = f(X)$. The inclusion map of a subset of Y is the restriction to that subset of the identity map on Y.

Here is a simple but useful example of a function. Consider any two sets X and Y, and define a function f from $X \times Y$ onto X by writing $f(x, y) = x$. (The purist will have noted that we should have written $f((x, y))$ instead of $f(x, y)$, but nobody ever does.) The function f is called the *projection* from $X \times Y$ onto X; if, similarly, $g(x, y) = y$, then g is the projection from $X \times Y$ onto Y. The terminology here is at variance with an earlier one, but not too badly. If $R = X \times Y$, then what was earlier called the projection of R onto the first coordinate is, in the present language, the range of the projection f.

A more complicated and correspondingly more valuable example of a function can be obtained as follows. Suppose R is an equivalence relation in X, and let f be the function from X onto X/R defined by $f(x) = x/R$. The function f is sometimes called the *canonical map* from X to X/R.

If f is an arbitrary function, from X onto Y, then there is a natural way of defining an equivalence relation R in X; write $a\,R\,b$ (where a and b are in X) in case $f(a) = f(b)$. For each element y of Y, let $g(y)$ be the set of all those elements x in X for which $f(x) = y$. The definition of R implies that $g(y)$ is, for each y, an equivalence class of the relation R; in other words, g is a function from Y onto the set X/R of all equivalence classes of R. The function g has the following special property: if u and v are distinct elements of Y, then $g(u)$ and $g(v)$ are distinct elements of X/R. A function that always maps distinct elements onto distinct elements is called *one-to-one* (usually a *one-to-one correspondence*). Among the examples above the inclusion maps are one-to-one, but, except in some trivial special cases, the projections are not. (Exercise: what special cases?)

To introduce the next aspect of the elementary theory of functions we must digress for a moment and anticipate a tiny fragment of our ultimate definition of natural numbers. We shall not find it necessary to define all the natural numbers now; all we need is the first three of them. Since this is not the appropriate occasion for lengthy heuristic preliminaries, we shall proceed directly to the definition, even at the risk of temporarily shocking or worrying some readers. Here it is: we define 0, 1, and 2 by writing

$$0 = \varnothing, \quad 1 = \{\varnothing\}, \quad \text{and} \quad 2 = \{\varnothing, \{\varnothing\}\}.$$

In other words, 0 is empty, 1 is the singleton $\{0\}$, and 2 is the pair $\{0, 1\}$.

Observe that there is some method in this apparent madness; the number of elements in the sets 0, 1, or 2 (in the ordinary everyday sense of the word) is, respectively, zero, one, or two.

If A is a subset of a set X, the *characteristic function* of A is the function χ from X to 2 such that $\chi(x) = 1$ or 0 according as $x \in A$ or $x \in X - A$. The dependence of the characteristic function of A on the set A may be indicated by writing χ_A instead of χ. The function that assigns to each subset A of X (that is, to each element of $\mathcal{P}(X)$) the characteristic function of A (that is, an element of 2^X) is a one-to-one correspondence between $\mathcal{P}(X)$ and 2^X. (Parenthetically: instead of the phrase "the function that assigns to each A in $\mathcal{P}(X)$ the element χ_A in 2^X" it is customary to use the abbreviation "the function $A \rightarrow \chi_A$." In this language, the projection from $X \times Y$ onto X, for instance, may be called the function $(x, y) \rightarrow x$, and the canonical map from a set X with a relation R onto X/R may be called the function $x \rightarrow x/R$.)

EXERCISE. (i) $Y^\varnothing$ has exactly one element, namely $\varnothing$, whether Y is empty or not, and (ii) if X is not empty, then $\varnothing^X$ is empty.

SECTION 9

FAMILIES

There are occasions when the range of a function is deemed to be more important than the function itself. When that is the case, both the terminology and the notation undergo radical alterations. Suppose, for instance, that x is a function from a set I to a set X. (The very choice of letters indicates that something strange is afoot.) An element of the domain I is called an *index*, I is called the *index set*, the range of the function is called an *indexed set*, the function itself is called a *family*, and the value of the function x at an index i, called a *term* of the family, is denoted by x_i. (This terminology is not absolutely established, but it is one of the standard choices among related slight variants; in the sequel it and it alone will be used.) An unacceptable but generally accepted way of communicating the notation and indicating the emphasis is to speak of a family $\{x_i\}$ in X, or of a family $\{x_i\}$ of whatever the elements of X may be; when necessary, the index set I is indicated by some such parenthetical expression as $(i \in I)$. Thus, for instance, the phrase "a family $\{A_i\}$ of subsets of X" is usually understood to refer to a function A, from some set I of indices, into $\mathcal{P}(X)$.

If $\{A_i\}$ is a family of subsets of X, the union of the range of the family is called the union of the family $\{A_i\}$, or the union of the sets A_i; the standard notation for it is

$$\bigcup_{i \in I} A_i \quad \text{or} \quad \bigcup_i A_i,$$

according as it is or is not important to emphasize the index set I. It follows immediately from the definition of unions that $x \in \bigcup_i A_i$ if and only if x belongs to A_i for at least one i. If $I = 2$, so that the range of the family $\{A_i\}$ is the unordered pair $\{A_0, A_1\}$, then $\bigcup_i A_i = A_0 \cup A_1$. Observe that there is no loss of generality in considering families of sets instead of arbitrary collections of sets; every collection of sets is the range

34

of some family. If, indeed, $\mathcal{C}$ is a collection of sets, let $\mathcal{C}$ itself play the role of the index set, and consider the identity mapping on $\mathcal{C}$ in the role of the family.

The algebraic laws satisfied by the operation of union for pairs can be generalized to arbitrary unions. Suppose, for instance, that $\{I_j\}$ is a family of sets with domain J, say; write $K = \bigcup_j I_j$, and let $\{A_k\}$ be a family of sets with domain K. It is then not difficult to prove that

$$\bigcup_{k \in K} A_k = \bigcup_{j \in J} \left(\bigcup_{i \in I_j} A_i \right);$$

this is the generalized version of the associative law for unions. Exercise: formulate and prove a generalized version of the commutative law.

An empty union makes sense (and is empty), but an empty intersection does not make sense. Except for this triviality, the terminology and notation for intersections parallels that for unions in every respect. Thus, for instance, if $\{A_i\}$ is a non-empty family of sets, the intersection of the range of the family is called the intersection of the family $\{A_i\}$, or the intersection of the sets A_i; the standard notation for it is

$$\bigcap_{i \in I} A_i \quad \text{or} \quad \bigcap_i A_i,$$

according as it is or is not important to emphasize the index set I. (By a "non-empty family" we mean a family whose domain I is not empty.) It follows immediately from the definition of intersections that if $I \neq \varnothing$, then a necessary and sufficient condition that x belong to $\bigcap_i A_i$ is that x belong to A_i for all i.

The generalized commutative and associative laws for intersections can be formulated and proved the same way as for unions, or, alternatively, De Morgan's laws can be used to derive them from the facts for unions. This is almost obvious, and, therefore, it is not of much interest. The interesting algebraic identities are the ones that involve both unions and intersections. Thus, for instance, if $\{A_i\}$ is a family of subsets of X and $B \subset X$, then

$$B \cap \bigcup_i A_i = \bigcup_i (B \cap A_i)$$

and

$$B \cup \bigcap_i A_i = \bigcap_i (B \cup A_i);$$

these equations are a mild generalization of the distributive laws.

Exercise. If both $\{A_i\}$ and $\{B_j\}$ are families of sets, then

$$\left(\bigcup_i A_i \right) \cap \left(\bigcup_j B_j \right) = \bigcup_{i,j} (A_i \cap B_j)$$

and

$$\left(\bigcap_i A_i \right) \cup \left(\bigcap_j B_j \right) = \bigcap_{i,j} (A_i \cup B_j).$$

Explanation of notation: a symbol such as $\bigcup_{i,j}$ is an abbreviation for $\bigcup_{(i,j)\,\epsilon\,I\,\times\,J}$.

The notation of families is the one normally used in generalizing the concept of Cartesian product. The Cartesian product of two sets X and Y was defined as the set of all ordered pairs (x, y) with x in X and y in Y. There is a natural one-to-one correspondence between this set and a certain set of families. Consider, indeed, any particular unordered pair $\{a, b\}$, with $a \neq b$, and consider the set Z of all families z, indexed by $\{a, b\}$, such that $z_a \,\epsilon\, X$ and $z_b \,\epsilon\, Y$. If the function f from Z to $X \times Y$ is defined by $f(z) = (z_a, z_b)$, then f is the promised one-to-one correspondence. The difference between Z and $X \times Y$ is merely a matter of notation. The generalization of Cartesian products generalizes Z rather than $X \times Y$ itself. (As a consequence there is a little terminological friction in the passage from the special case to the general. There is no help for it; that is how mathematical language is in fact used nowadays.) The generalization is now straightforward. If $\{X_i\}$ is a family of sets $(i \,\epsilon\, I)$, the *Cartesian product* of the family is, by definition, the set of all families $\{x_i\}$ with $x_i \,\epsilon\, X_i$ for each i in I. There are several symbols for the Cartesian product in more or less current usage; in this book we shall denote it by

$$\mathop{\times}_{i\,\epsilon\,I} X_i \quad \text{or} \quad \mathop{\times}_i X_i.$$

It is clear that if every X_i is equal to one and the same set X, then $\mathop{\times}_i X_i = X^I$. If I is a pair $\{a, b\}$, with $a \neq b$, then it is customary to identify $\mathop{\times}_{i\,\epsilon\,I} X_i$ with the Cartesian product $X_a \times X_b$ as defined earlier, and if I is a singleton $\{a\}$, then, similarly, we identify $\mathop{\times}_{i\,\epsilon\,I} X_i$ with X_a itself. *Ordered triples, ordered quadruples,* etc., may be defined as families whose index sets are unordered triples, quadruples, etc.

Suppose that $\{X_i\}$ is a family of sets $(i \,\epsilon\, I)$ and let X be its Cartesian product. If J is a subset of I, then to each element of X there corresponds in a natural way an element of the partial Cartesian product $\mathop{\times}_{i\,\epsilon\,J} X_i$. To define the correspondence, recall that each element x of X is itself a family $\{x_i\}$, that is, in the last analysis, a function on I; the corresponding element, say y, of $\mathop{\times}_{i\,\epsilon\,J} X_i$ is obtained by simply restricting that function to J. Explicitly, we write $y_i = x_i$ whenever $i \,\epsilon\, J$. The correspondence $x \rightarrow y$ is called the projection from X onto $\mathop{\times}_{i\,\epsilon\,J} X_i$; we shall temporarily denote it by f_J. If, in particular, J is a singleton, say $J = \{j\}$, then we shall write f_j (instead of $f_{\{j\}}$) for f_J. The word "projection" has a multiple use; if $x \,\epsilon\, X$, the value of f_j at x, that is x_j, is also called the projection of x onto X_j, or, alternatively, the *j-coordinate* of x. A function on a Cartesian product such as X is called a function of *several variables*, and, in particular, a function on a Cartesian product $X_a \times X_b$ is called a function of two variables.

EXERCISE. Prove that $(\bigcup_i A_i) \times (\bigcup_j B_j) = \bigcup_{i,j} (A_i \times B_j)$, and that the same equation holds for intersections (provided that the domains of the families involved are not empty). Prove also (with appropriate provisos about empty families) that $\bigcap_i X_i \subset X_j \subset \bigcup_i X_i$ for each index j and that intersection and union can in fact be characterized as the extreme solutions of these inclusions. This means that if $X_j \subset Y$ for each index j, then $\bigcup_i X_i \subset Y$, and that $\bigcup_i X_i$ is the only set satisfying this minimality condition; the formulation for intersections is similar.

SECTION 10

INVERSES AND COMPOSITES

Associated with every function f, from X to Y, say, there is a function from $\mathcal{P}(X)$ to $\mathcal{P}(Y)$, namely the function (frequently called f also) that assigns to each subset A of X the image subset $f(A)$ of Y. The algebraic behavior of the mapping $A \to f(A)$ leaves something to be desired. It is true that if $\{A_i\}$ is a family of subsets of X, then $f(\bigcup_i A_i) = \bigcup_i f(A_i)$ (proof?), but the corresponding equation for intersections is false in general (example?), and the connection between images and complements is equally unsatisfactory.

A correspondence between the elements of X and the elements of Y does always induce a well-behaved correspondence between the subsets of X and the subsets of Y, not forward, by the formation of images, but backward, by the formation of inverse images. Given a function f from X to Y, let f^{-1}, the *inverse* of f, be the function from $\mathcal{P}(Y)$ to $\mathcal{P}(X)$ such that if $B \subset Y$, then

$$f^{-1}(B) = \{x \in X : f(x) \in B\}.$$

In words: $f^{-1}(B)$ consists of exactly those elements of X that f maps into B; the set $f^{-1}(B)$ is called the *inverse image* of B under f. A necessary and sufficient condition that f map X onto Y is that the inverse image under f of each non-empty subset of Y be a non-empty subset of X. (Proof?) A necessary and sufficient condition that f be one-to-one is that the inverse image under f of each singleton in the range of f be a singleton in X.

If the last condition is satisfied, then the symbol f^{-1} is frequently assigned a second interpretation, namely as the function whose domain is the range of f, and whose value for each y in the range of f is the unique x in X for which $f(x) = y$. In other words, for one-to-one functions f we may write $f^{-1}(y) = x$ if and only if $f(x) = y$. This use of the notation is

38

mildly inconsistent with our first interpretation of f^{-1}, but the double meaning is not likely to lead to any confusion.

The connection between images and inverse images is worth a moment's consideration.

If $B \subset Y$, then
$$f(f^{-1}(B)) \subset B.$$

Proof. If $y \in f(f^{-1}(B))$, then $y = f(x)$ for some x in $f^{-1}(B)$; this means that $y = f(x)$ and $f(x) \in B$, and therefore $y \in B$.

If f maps X onto Y, then
$$f(f^{-1}(B)) = B.$$

Proof. If $y \in B$, then $y = f(x)$ for some x in X, and therefore for some x in $f^{-1}(B)$; this means that $y \in f(f^{-1}(B))$.

If $A \subset X$, then
$$A \subset f^{-1}(f(A)).$$

Proof. If $x \in A$, then $f(x) \in f(A)$; this means that $x \in f^{-1}(f(A))$.

If f is one-to-one, then
$$A = f^{-1}(f(A)).$$

Proof. If $x \in f^{-1}(f(A))$, then $f(x) \in f(A)$, and therefore $f(x) = f(u)$ for some u in A; this implies that $x = u$ and hence that $x \in A$.

The algebraic behavior of f^{-1} is unexceptionable. If $\{B_i\}$ is a family of subsets of Y, then
$$f^{-1}(\bigcup_i B_i) = \bigcup_i f^{-1}(B_i)$$
and
$$f^{-1}(\bigcap_i B_i) = \bigcap_i f^{-1}(B_i).$$

The proofs are straightforward. If, for instance, $x \in f^{-1}(\bigcap_i B_i)$, then $f(x) \in B_i$ for all i, so that $x \in f^{-1}(B_i)$ for all i, and therefore $x \in \bigcap_i f^{-1}(B_i)$; all the steps in this argument are reversible. The formation of inverse images commutes with complementation also; i.e.,
$$f^{-1}(Y - B) = X - f^{-1}(B)$$

for each subset B of Y. Indeed: if $x \in f^{-1}(Y - B)$, then $f(x) \in Y - B$, so that $x \in' f^{-1}(B)$, and therefore $x \in X - f^{-1}(B)$; the steps are reversible. (Observe that the last equation is indeed a kind of commutative law: it says that complementation followed by inversion is the same as inversion followed by complementation.)

The discussion of inverses shows that what a function does can in a certain sense be undone; the next thing we shall see is that what two functions do can sometimes be done in one step. If, to be explicit, f is a function from X to Y and g is a function from Y to Z, then every element in the range of f belongs to the domain of g, and, consequently, $g(f(x))$ makes sense for each x in X. The function h from X to Z, defined by $h(x) = g(f(x))$ is called the *composite* of the functions f and g; it is denoted by $g \circ f$ or, more simply, by gf. (Since we shall not have occasion to consider any other kind of multiplication for functions, in this book we shall use the latter, simpler notation only.)

Observe that the order of events is important in the theory of functional composition. In order that gf be defined, the range of f must be included in the domain of g, and this can happen without it necessarily happening in the other direction at the same time. Even if both fg and gf are defined, which happens if, for instance, f maps X into Y and g maps Y into X, the functions fg and gf need not be the same; in other words, functional composition is not necessarily commutative.

Functional composition may not be commutative, but it is always associative. If f maps X into Y, if g maps Y into Z, and if h maps Z into U, then we can form the composite of h with gf and the composite of hg with f; it is a simple exercise to show that the result is the same in either case.

The connection between inversion and composition is important; something like it crops up all over mathematics. If f maps X into Y and g maps Y into Z, then f^{-1} maps $\mathcal{P}(Y)$ into $\mathcal{P}(X)$ and g^{-1} maps $\mathcal{P}(Z)$ into $\mathcal{P}(Y)$. In this situation, the composites that are formable are gf and $f^{-1}g^{-1}$; the assertion is that the latter is the inverse of the former. Proof: if $x \in (gf)^{-1}(C)$, where $x \in X$ and $C \subset Z$, then $g(f(x)) \in C$, so that $f(x) \in g^{-1}(C)$, and therefore $x \in f^{-1}(g^{-1}(C))$; the steps of the argument are reversible.

Inversion and composition for functions are special cases of similar operations for relations. Thus, in particular, associated with every relation R from X to Y there is the *inverse* (or *converse*) relation R^{-1} from Y to X; by definition $y\,R^{-1}\,x$ means that $x\,R\,y$. Example: if R is the relation of belonging, from X to $\mathcal{P}(X)$, then R^{-1} is the relation of containing, from $\mathcal{P}(X)$ to X. It is an immediate consequence of the definitions involved that $\operatorname{dom} R^{-1} = \operatorname{ran} R$ and $\operatorname{ran} R^{-1} = \operatorname{dom} R$. If the relation R is a function, then the equivalent assertions $x\,R\,y$ and $y\,R^{-1}\,x$ can be written in the equivalent forms $R(x) = y$ and $x \in R^{-1}(\{y\})$.

Because of difficulties with commutativity, the generalization of functional composition has to be handled with care. The composite of the relations R and S is defined in case R is a relation from X to Y and S is a relation from Y to Z. The composite relation T, from X to Z, is denoted by $S \circ R$, or, simply, by SR; it is defined so that $x\ T\ z$ if and only if there exists an element y in Y such that $x\ R\ y$ and $y\ S\ z$. For an instructive example, let R mean "son" and let S mean "brother" in the set of human males, say. In other words, $x\ R\ y$ means that x is a son of y, and $y\ S\ z$ means that y is a brother of z. In this case the composite relation SR means "nephew." (Query: what do R^{-1}, S^{-1}, RS, and $R^{-1}S^{-1}$ mean?) If both R and S are functions, then $x\ R\ y$ and $y\ S\ z$ can be rewritten as $R(x) = y$ and $S(y) = z$, respectively. It follows that $S(R(x)) = z$ if and only if $x\ T\ z$, so that functional composition is indeed a special case of what is sometimes called the *relative product*.

The algebraic properties of inversion and composition are the same for relations as for functions. Thus, in particular, composition is commutative by accident only, but it is always associative, and it is always connected with inversion via the equation $(SR)^{-1} = R^{-1}S^{-1}$. (Proofs?)

The algebra of relations provides some amusing formulas. Suppose that, temporarily, we consider relations in one set X only, and, in particular, let I be the relation of equality in X (which is the same as the identity mapping on X). The relation I acts as a multiplicative unit; this means that $IR = RI = R$ for every relation R in X. Query: is there a connection among I, RR^{-1}, and $R^{-1}R$? The three defining properties of an equivalence relation can be formulated in algebraic terms as follows: reflexivity means $I \subset R$, symmetry means $R \subset R^{-1}$, and transitivity means $RR \subset R$.

EXERCISE. (Assume in each case that f is a function from X to Y.) (i) If g is a function from Y to X such that gf is the identity on X, then f is one-to-one and g maps Y onto X. (ii) A necessary and sufficient condition that $f(A \cap B) = f(A) \cap f(B)$ for all subsets A and B of X is that f be one-to-one. (iii) A necessary and sufficient condition that $f(X - A) \subset Y - f(A)$ for all subsets A of X is that f be one-to-one. (iv) A necessary and sufficient condition that $Y - f(A) \subset f(X - A)$ for all subsets A of X is that f map X onto Y.

SECTION 11

NUMBERS

How much is two? How, more generally, are we to define numbers? To prepare for the answer, let us consider a set X and let us form the collection P of all unordered pairs $\{a, b\}$, with a in X, b in X, and $a \neq b$. It seems clear that all the sets in the collection P have a property in common, namely the property of consisting of two elements. It is tempting to try to define "twoness" as the common property of all the sets in the collection P, but the temptation must be resisted; such a definition is, after all, mathematical nonsense. What is a "property"? How do we know that there is only one property in common to all the sets in P?

After some cogitation we might hit upon a way of saving the idea behind the proposed definition without using vague expressions such as "the common property." It is ubiquitous mathematical practice to identify a property with a set, namely with the set of all objects that possess the property; why not do it here? Why not, in other words, define "two" as the set P? Something like this is done at times, but it is not completely satisfying. The trouble is that our present modified proposal depends on P, and hence ultimately on X. At best the proposal defines twoness for subsets of X; it gives no hint as to when we may attribute twoness to a set that is not included in X.

There are two ways out. One way is to abandon the restriction to a particular set and to consider instead all possible unordered pairs $\{a, b\}$ with $a \neq b$. These unordered pairs do not constitute a set; in order to base the definition of "two" on them, the entire theory under consideration would have to be extended to include the "unsets" (classes) of another theory. This can be done, but it will not be done here; we shall follow a different route.

How would a mathematician define a meter? The procedure analogous to the one sketched above would involve the following two steps. First, select an object that is one of the intended models of the concept being defined—an object, in other words, such that on intuitive or practical grounds it deserves to be called one meter long if anything does. Second, form the set of all objects in the universe that are of the same length as the selected one (note that this does not depend on knowing what a meter is), and define a meter as the set so formed.

How in fact is a meter defined? The example was chosen so that the answer to this question should suggest an approach to the definition of numbers. The point is that in the customary definition of a meter the second step is omitted. By a more or less arbitrary convention an object is selected and its length is called a meter. If the definition is accused of circularity (what does "length" mean?), it can easily be converted into an unexceptionable demonstrative definition; there is after all nothing to stop us from defining a meter as equal to the selected object. If this demonstrative approach is adopted, it is just as easy to explain as before when "one-meter-ness" shall be attributed to some other object, namely, just in case the new object has the same length as the selected standard. We comment again that to determine whether two objects have the same length depends on a simple act of comparison only, and does not depend on having a precise definition of length.

Motivated by the considerations described above, we have earlier defined 2 as some particular set with (intuitively speaking) exactly two elements. How was that standard set selected? How should other such standard sets for other numbers be selected? There is no compelling mathematical reason for preferring one answer to this question to another; the whole thing is largely a matter of taste. The selection should presumably be guided by considerations of simplicity and economy. To motivate the particular selection that is usually made, suppose that a number, say 7, has already been defined as a set (with seven elements). How, in this case, should we define 8? Where, in other words, can we find a set consisting of exactly eight elements? We can find seven elements in the set 7; what shall we use as an eighth to adjoin to them? A reasonable answer to the last question is the number (set) 7 itself; the proposal is to define 8 to be the set consisting of the seven elements of 7, together with 7. Note that according to this proposal each number will be equal to the set of its own predecessors.

The preceding paragraph motivates a set-theoretic construction that makes sense for every set, but that is of interest in the construction of numbers only. For every set x we define the *successor* x^+ of x to be the set obtained by adjoining x to the elements of x; in other words,

$$x^+ = x \cup \{x\}.$$

(The successor of x is frequently denoted by x'.)

We are now ready to define the natural numbers. In defining 0 to be a set with zero elements, we have no choice; we must write (as we did)

$$0 = \varnothing.$$

If every natural number is to be equal to the set of its predecessors, we have no choice in defining 1, or 2, or 3 either; we must write

$$1 = 0^+ \,(= \{0\}),$$

$$2 = 1^+ \,(= \{0, 1\}),$$

$$3 = 2^+ \,(= \{0, 1, 2\}),$$

etc. The "etc." means that we hereby adopt the usual notation, and, in what follows, we shall feel free to use numerals such as "4" or "956" without any further explanation or apology.

From what has been said so far it does not follow that the construction of successors can be carried out ad infinitum within one and the same set. What we need is a new set-theoretic principle.

Axiom of infinity. *There exists a set containing 0 and containing the successor of each of its elements.*

The reason for the name of the axiom should be clear. We have not yet given a precise definition of infinity, but it seems reasonable that sets such as the ones that the axiom of infinity describes deserve to be called infinite.

We shall say, temporarily, that a set A is a *successor set* if $0 \,\epsilon\, A$ and if $x^+ \,\epsilon\, A$ whenever $x \,\epsilon\, A$. In this language the axiom of infinity simply says that there exists a successor set A. Since the intersection of every (non-empty) family of successor sets is a successor set itself (proof?), the intersection of all the successor sets included in A is a successor set ω. The set ω is a subset of every successor set. If, indeed, B is an arbitrary successor set, then so is $A \cap B$. Since $A \cap B \subset A$, the set $A \cap B$ is one of the sets that entered into the definition of ω; it follows that $\omega \subset A \cap B$, and, consequently, that $\omega \subset B$. The minimality property so established uniquely characterizes ω; the axiom of extension guarantees that there can be only one successor set that is included in every other successor set. A *natural number* is, by definition, an element of the minimal successor set ω. This definition of natural numbers is the rigorous counterpart of the intuitive description according to which they consist of 0, 1, 2, 3, "and so on." Incidentally, the symbol we are using for the set of all natural numbers (ω) has a plurality of the votes of the writers on the subject, but nothing like a clear majority. In this book that symbol will be used systematically and exclusively in the sense defined above.

The slight feeling of discomfort that the reader may experience in connection with the definition of natural numbers is quite common and in most cases temporary. The trouble is that here, as once before (in the definition of ordered pairs), the object defined has some irrelevant structure, which seems to get in the way (but is in fact harmless). We want to be told that the successor of 7 is 8, but to be told that 7 is a subset of 8 or that 7 is an element of 8 is disturbing. We shall make use of this superstructure of natural numbers just long enough to derive their most important natural properties; after that the superstructure may safely be forgotten.

A family $\{x_i\}$ whose index set is either a natural number or else the set of all natural numbers is called a *sequence* (*finite* or *infinite*, respectively). If $\{A_i\}$ is a sequence of sets, where the index set is the natural number n^+, then the union of the sequence is denoted by

$$\bigcup_{i=0}^{n} A_i \quad \text{or} \quad A_0 \cup \cdots \cup A_n.$$

If the index set is ω, the notation is

$$\bigcup_{i=0}^{\infty} A_i \quad \text{or} \quad A_0 \cup A_1 \cup A_2 \cup \cdots.$$

Intersections and Cartesian products of sequences are denoted similarly by

$$\bigcap_{i=0}^{n} A_i, \quad A_0 \cap \cdots \cap A_n,$$

$$\mathsf{X}_{i=0}^{n} A_i, \quad A_0 \times \cdots \times A_n,$$

and

$$\bigcap_{i=0}^{\infty} A_i, \quad A_0 \cap A_1 \cap A_2 \cap \cdots,$$

$$\mathsf{X}_{i=0}^{\infty} A_i, \quad A_0 \times A_1 \times A_2 \times \cdots.$$

The word "sequence" is used in a few different ways in the mathematical literature, but the differences among them are more notational than conceptual. The most common alternative starts at 1 instead of 0; in other words, it refers to a family whose index set is $\omega - \{0\}$ instead of ω.

SECTION 12

THE PEANO AXIOMS

We enter now into a minor digression. The purpose of the digression is to make fleeting contact with the arithmetic theory of natural numbers. From the set-theoretic point of view this is a pleasant luxury.

The most important thing we know about the set ω of all natural numbers is that it is the unique successor set that is a subset of every successor set. To say that ω is a successor set means that

(I) $\qquad\qquad\qquad\qquad 0 \,\epsilon\, \omega$

(where, of course, $0 = \varnothing$), and that

(II) $\qquad\qquad\qquad$ *if $n \,\epsilon\, \omega$, then $n^+ \,\epsilon\, \omega$*

(where $n^+ = n \cup \{n\}$). The minimality property of ω can be expressed by saying that if a subset S of ω is a successor set, then $S = \omega$. Alternatively, and in more primitive terms,

(III) $\quad$ *if $S \subset \omega$, if $0 \,\epsilon\, S$, and if $n^+ \,\epsilon\, S$ whenever $n \,\epsilon\, S$, then $S = \omega$.*

Property (III) is known as the **principle of mathematical induction**. We shall now add to this list of properties of ω two others:

(IV) $\qquad\qquad\qquad n^+ \neq 0$ *for all n in ω,*

and

(V) $\qquad$ *if n and m are in ω, and if $n^+ = m^+$, then $n = m$.*

The proof of (IV) is trivial; since n^+ always contains n, and since 0 is empty, it is clear that n^+ is different from 0. The proof of (V) is not trivial; it depends on a couple of auxiliary propositions. The first one asserts that something that ought not to happen indeed does not happen. Even

46

if the considerations that the proof involves seem to be pathological and foreign to the arithmetic spirit that we expect to see in the theory of natural numbers, the end justifies the means. The second proposition refers to behavior that is quite similar to the one just excluded. This time, however, the apparently artificial considerations end in an affirmative result: something mildly surprising always does happen. The statements are as follows: (i) *no natural number is a subset of any of its elements*, and (ii) *every element of a natural number is a subset of it*. Sometimes a set with the property that it includes ($\subset$) everything that it contains (ϵ) is called a *transitive* set. More precisely, to say that E is transitive means that if $x \epsilon y$ and $y \epsilon E$, then $x \epsilon E$. (Recall the slightly different use of the word that we encountered in the theory of relations.) In this language, (ii) says that every natural number is transitive.

The proof of (i) is a typical application of the principle of mathematical induction. Let S be the set of all those natural numbers n that are not included in any of their elements. (Explicitly: $n \epsilon S$ if and only if $n \epsilon \omega$ and n is not a subset of x for any x in n.) Since 0 is not a subset of any of its elements, it follows that $0 \epsilon S$. Suppose now that $n \epsilon S$. Since n is a subset of n, we may infer that n is not an element of n, and hence that n^+ is not a subset of n. What can n^+ be a subset of? If $n^+ \subset x$, then $n \subset x$, and therefore (since $n \epsilon S$) $x \epsilon' n$. It follows that n^+ cannot be a subset of n, and n^+ cannot be a subset of any element of n. This means that n^+ cannot be a subset of any element of n^+, and hence that $n^+ \epsilon S$. The desired conclusion (i) is now a consequence of (III).

The proof of (ii) is also inductive. This time let S be the set of all transitive natural numbers. (Explicitly: $n \epsilon S$ if and only if $n \epsilon \omega$ and x is a subset of n for every x in n.) The requirement that $0 \epsilon S$ is vacuously satisfied. Suppose now that $n \epsilon S$. If $x \epsilon n^+$, then either $x \epsilon n$ or $x = n$. In the first case $x \subset n$ (since $n \epsilon S$) and therefore $x \subset n^+$; in the second case $x \subset n^+$ for even more trivial reasons. It follows that every element of n^+ is a subset of n^+, or, in other words, that $n^+ \epsilon S$. The desired conclusion (ii) is a consequence of (III).

We are now ready to prove (V). Suppose indeed that n and m are natural numbers and that $n^+ = m^+$. Since $n \epsilon n^+$, it follows that $n \epsilon m^+$, and hence that either $n \epsilon m$ or $n = m$. Similarly, either $m \epsilon n$ or $m = n$. If $n \neq m$, then we must have $n \epsilon m$ and $m \epsilon n$. Since, by (ii), n is transitive, it follows that $n \epsilon n$. Since, however, $n \subset n$, this contradicts (i), and the proof is complete.

The assertions (I)–(V) are known as the Peano axioms; they used to be considered as the fountainhead of all mathematical knowledge. From them (together with the set-theoretic principles we have already met) it is possible to define integers, rational numbers, real numbers, and complex numbers, and to derive their usual arithmetic and analytic properties. Such a program is not within the scope of this book; the interested reader should have no difficulty in locating and studying it elsewhere.

Induction is often used not only to prove things but also to define things. Suppose, to be specific, that f is a function from a set X into the same set X, and suppose that a is an element of X. It seems natural to try to define an infinite sequence $\{u(n)\}$ of elements of X (that is, a function u from ω to X) in some such way as this: write $u(0) = a$, $u(1) = f(u(0))$, $u(2) = f(u(1))$, and so on. If the would-be definer were pressed to explain the "and so on," he might lean on induction. What it all means, he might say, is that we define $u(0)$ as a, and then, inductively, we define $u(n^+)$ as $f(u(n))$ for every n. This may sound plausible, but, as justification for an existential assertion, it is insufficient. The principle of mathematical induction does indeed prove, easily, that there can be at most one function satisfying all the stated conditions, but it does not establish the existence of such a function. What is needed is the following result.

Recursion theorem. *If a is an element of a set X, and if f is a function from X into X, then there exists a function u from ω into X such that $u(0) = a$ and such that $u(n^+) = f(u(n))$ for all n in ω.*

PROOF. Recall that a function from ω to X is a certain kind of subset of $\omega \times X$; we shall construct u explicitly as a set of ordered pairs. Consider, for this purpose, the collection $\mathcal{C}$ of all those subsets A of $\omega \times X$ for which $(0, a) \epsilon A$ and for which $(n^+, f(x)) \epsilon A$ whenever $(n, x) \epsilon A$. Since $\omega \times X$ has these properties, the collection $\mathcal{C}$ is not empty. We may, therefore, form the intersection u of all the sets of the collection $\mathcal{C}$. Since it is easy to see that u itself belongs to $\mathcal{C}$, it remains only to prove that u is a function. We are to prove, in other words, that for each natural number n there exists at most one element x of X such that $(n, x) \epsilon u$. (Explicitly: if both (n, x) and (n, y) belong to u, then $x = y$.) The proof is inductive. Let S be the set of all those natural numbers n for which it is indeed true that $(n, x) \epsilon u$ for at most one x. We shall prove that $0 \epsilon S$ and that if $n \epsilon S$, then $n^+ \epsilon S$.

Does 0 belong to S? If not, then $(0, b) \epsilon u$ for some b distinct from a. Consider, in this case, the set $u - \{(0, b)\}$. Observe that this diminished set still contains $(0, a)$ (since $a \neq b$), and that if the diminished set contains (n, x), then it contains $(n^+, f(x))$ also. The reason for the second assertion is that since $n^+ \neq 0$, the discarded element is not equal to

$(n^+, f(x))$. In other words, $u - \{(0, b)\} \, \epsilon \, \mathbb{C}$. This contradicts the fact that u is the smallest set in $\mathbb{C}$, and we may conclude that $0 \, \epsilon \, S$.

Suppose now that $n \, \epsilon \, S$; this means that there exists a unique element x in X such that $(n, x) \, \epsilon \, u$. Since $(n, x) \, \epsilon \, u$, it follows that $(n^+, f(x)) \, \epsilon \, u$. If n^+ does not belong to S, then $(n^+, y) \, \epsilon \, u$ for some y different from $f(x)$. Consider, in this case, the set $u - \{(n^+, y)\}$. Observe that this diminished set contains $(0, a)$ (since $n^+ \neq 0$), and that if the diminished set contains (m, t), say, then it contains $(m^+, f(t))$ also. Indeed, if $m = n$, then t must be x, and the reason the diminished set contains $(n^+, f(x))$ is that $f(x) \neq y$; if, on the other hand, $m \neq n$, then the reason the diminished set contains $(m^+, f(t))$ is that $m^+ \neq n^+$. In other words, $u - \{(n^+, y)\} \, \epsilon \, \mathbb{C}$. This again contradicts the fact that u is the smallest set in $\mathbb{C}$, and we may conclude that $n^+ \, \epsilon \, S$.

The proof of the recursion theorem is complete. An application of the recursion theorem is called *definition by induction*.

EXERCISE. Prove that if n is a natural number, then $n \neq n^+$; if $n \neq 0$, then $n = m^+$ for some natural number m. Prove that ω is transitive. Prove that if E is a non-empty subset of some natural number, then there exists an element k in E such that $k \, \epsilon \, m$ whenever m is an element of E distinct from k.

SECTION 13

ARITHMETIC

The introduction of addition for natural numbers is a typical example of definition by induction. Indeed, it follows from the recursion theorem that for each natural number m there exists a function s_m from ω to ω such that $s_m(0) = m$ and such that $s_m(n^+) = (s_m(n))^+$ for every natural number n; the value $s_m(n)$ is, by definition, the *sum* $m + n$. The general arithmetic properties of addition are proved by repeated applications of the principle of mathematical induction. Thus, for instance, addition is associative. This means that

$$(k + m) + n = k + (m + n)$$

whenever k, m, and n are natural numbers. The proof goes by induction on n as follows. Since $(k + m) + 0 = k + m$ and $k + (m + 0) = k + m$, the equation is true if $n = 0$. If the equation is true for n, then $(k + m) + n^+ = ((k + m) + n)^+$ (by definition) $= (k + (m + n))^+$ (by the induction hypothesis) $= k + (m + n)^+$ (again by the definition of addition) $= k + (m + n^+)$ (ditto), and the argument is complete. The proof that addition is commutative (i.e., $m + n = n + m$ for all m and n) is a little tricky; a straightforward attack might fail. The trick is to prove, by induction on n, that (i) $0 + n = n$ and (ii) $m^+ + n = (m + n)^+$, and then to prove the desired commutativity equation by induction on m, via (i) and (ii).

Similar techniques are applied in the definitions of products and exponents and in the derivations of their basic arithmetic properties. To define multiplication, apply the recursion theorem to produce functions p_m such that $p_m(0) = 0$ and such that $p_m(n^+) = p_m(n) + m$ for every natural number n; then the value $p_m(n)$ is, by definition, the *product* $m \cdot n$. (The dot is frequently omitted.) Multiplication is associative and commutative; the

50

proofs are straightforward adaptations of the ones that worked for addition. The distributive law (i.e., the assertion that $k \cdot (m + n) = k \cdot m + k \cdot n$ whenever k, m, and n are natural numbers) is another easy consequence of the principle of mathematical induction. (Use induction on n.) Anyone who has worked through sums and products in this way should have no trouble with exponents. The recursion theorem yields functions e_m such that $e_m(0) = 1$ and such that $e_m(n^+) = e_m(n) \cdot m$ for every natural number n; the value $e_m(n)$ is, by definition, the *power m^n*. The discovery and establishment of the properties of powers, as well as the detailed proofs of the statements about products, can safely be left as exercises for the reader.

The next topic that deserves some attention is the theory of order in the set of natural numbers. For this purpose we proceed to examine with some care the question of which natural numbers belong to which others. Formally, we say that two natural numbers m and n are comparable if $m \,\epsilon\, n$, or $m = n$, or $n \,\epsilon\, m$. Assertion: two natural numbers are always comparable. The proof of this assertion consists of several steps; it will be convenient to introduce some notation. For each n in ω, write $S(n)$ for the set of all m in ω that are comparable with n, and let S be the set of all those n for which $S(n) = \omega$. In these terms, the assertion is that $S = \omega$. We begin the proof by showing that $S(0) = \omega$ (i.e., that $0 \,\epsilon\, S$). Clearly $S(0)$ contains 0. If $m \,\epsilon\, S(0)$, then, since $m \,\epsilon\, 0$ is impossible, either $m = 0$ (in which case $0 \,\epsilon\, m^+$), or $0 \,\epsilon\, m$ (in which case, again, $0 \,\epsilon\, m^+$). Hence, in all cases, if $m \,\epsilon\, S(0)$, then $m^+ \,\epsilon\, S(0)$; this proves that $S(0) = \omega$. We complete the proof by showing that if $S(n) = \omega$, then $S(n^+) = \omega$. The fact that $0 \,\epsilon\, S(n^+)$ is immediate (since $n^+ \,\epsilon\, S(0)$); it remains to prove that if $m \,\epsilon\, S(n^+)$, then $m^+ \,\epsilon\, S(n^+)$. Since $m \,\epsilon\, S(n^+)$, therefore either $n^+ \,\epsilon\, m$ (in which case $n^+ \,\epsilon\, m^+$), or $n^+ = m$ (ditto), or $m \,\epsilon\, n^+$. In the latter case, either $m = n$ (in which case $m^+ = n^+$), or $m \,\epsilon\, n$. The last case, in turn, splits according to the behavior of m^+ and n: since $m^+ \,\epsilon\, S(n)$, we must have either $n \,\epsilon\, m^+$, or $n = m^+$, or $m^+ \,\epsilon\, n$. The first possibility is incompatible with the present situation (i.e., with $m \,\epsilon\, n$). The reason is that if $n \,\epsilon\, m^+$, then either $n \,\epsilon\, m$ or $n = m$, so that, in any case, $n \subset m$, and we know that no natural number is a subset of one of its elements. Both the remaining possibilities imply that $m^+ \,\epsilon\, n^+$, and the proof is complete.

The preceding paragraph implies that if m and n are in ω, then at least one of the three possibilities ($m \,\epsilon\, n$, $m = n$, $n \,\epsilon\, m$) must hold; it is easy to see that, in fact, always exactly one of them holds. (The reason is another application of the fact that a natural number is not a subset of one of its elements.) Another consequence of the preceding paragraph is that if n and m are distinct natural numbers, then a necessary and sufficient condition that $m \,\epsilon\, n$ is that $m \subset n$. Indeed, the implication from $m \,\epsilon\, n$ to $m \subset n$ is just the transitivity of n. If, conversely, $m \subset n$ and $m \neq n$, then $n \,\epsilon\, m$ cannot happen (for then m would be a subset of one of its elements), and therefore $m \,\epsilon\, n$. If $m \,\epsilon\, n$, or if, equivalently, m is a proper subset of n, we shall write $m < n$ and we shall say that m is *less* than n. If m is known to be either less than n or else equal to n, we write $m \leq n$. Note that $\leq$ and $<$ are relations in ω. The former is reflexive, but the latter is not; neither is symmetric; both are transitive. If $m \leq n$ and $n \leq m$, then $m = n$.

EXERCISE. Prove that if $m < n$, then $m + k < n + k$, and prove that if $m < n$ and $k \neq 0$, then $m \cdot k < n \cdot k$. Prove that if E is a non-empty set of natural numbers, then there exists an element k in E such that $k \leq m$ for all m in E.

Two sets E and F (not necessarily subsets of ω) are called *equivalent*, in symbols $E \sim F$, if there exists a one-to-one correspondence between them. It is easy to verify that equivalence in this sense, for subsets of some particular set X, is an equivalence relation in the power set $\mathcal{P}(X)$.

Every proper subset of a natural number n is equivalent to some smaller natural number (i.e., to some element of n). The proof of this assertion is inductive. For $n = 0$ it is trivial. If it is true for n, and if E is a proper subset of n^+, then either E is a proper subset of n and the induction hypothesis applies, or $E = n$ and the result is trivial, or $n \,\epsilon\, E$. In the latter case, find a number k in n but not in E and define a function f on E by writing $f(i) = i$ when $i \neq n$ and $f(n) = k$. Clearly f is one-to-one and f maps E into n. It follows that the image of E under f is either equal to n or (by the induction hypothesis) equivalent to some element of n, and, consequently, E itself is always equivalent to some element of n^+.

It is a mildly shocking fact that a set can be equivalent to a proper subset of itself. If, for instance, a function f from ω to ω is defined by writing $f(n) = n^+$ for all n in ω, then f is a one-to-one correspondence between the set of all natural numbers and the proper subset consisting of the non-zero natural numbers. It is nice to know that even though the set of all natural numbers has this peculiar property, sanity prevails for each particular natural number. In other words, if $n \,\epsilon\, \omega$, then n is not equivalent to a proper subset of n. For $n = 0$ this is clear. Suppose now that it is true for n, and suppose that f is a one-to-one correspondence from n^+ to a proper subset E of n^+. If $n \,\epsilon'\, E$, then the restriction of f to n is a one-to-one correspondence between n and a proper subset of n, which contradicts the induction hypothesis. If $n \,\epsilon\, E$, then n is equivalent to $E - \{n\}$, so that, by the induction hypothesis, $n = E - \{n\}$. This implies that $E = n^+$, which contradicts the assumption that E is a proper subset of n^+.

A set E is called *finite* if it is equivalent to some natural number; otherwise E is *infinite*.

EXERCISE. Use this definition to prove that ω is infinite.

A set can be equivalent to at most one natural number. (Proof: we know that for any two distinct natural numbers one must be an element and therefore a proper subset of the other; it follows from the preceding paragraph that they cannot be equivalent.) We may infer that a finite set is never equivalent to a proper subset; in other words, as long as we stick to finite sets, the whole is always greater than any of its parts.

EXERCISE. Use this consequence of the definition of finiteness to prove that ω is infinite.

Since every subset of a natural number is equivalent to a natural number, it follows also that every subset of a finite set is finite.

The *number of elements* in a finite set E is, by definition, the unique natural number equivalent to E; we shall denote it by $\#(E)$. It is clear that if the correspondence between E and $\#(E)$ is restricted to the finite subsets of some set X, the result is a function from a subset of the power set $\mathcal{P}(X)$ to ω. This function is pleasantly related to the familiar set-theoretic relations and operations. Thus, for example, if E and F are finite sets such that $E \subset F$, then $\#(E) \leq \#(F)$. (The reason is that since $E \sim \#(E)$ and $F \sim \#(F)$, it follows that $\#(E)$ is equivalent to a subset of $\#(F)$.) Another example is the assertion that if E and F are finite sets, then $E \cup F$ is finite, and, moreover, if E and F are disjoint, then $\#(E \cup F) = \#(E) + \#(F)$. The crucial step in the proof is the fact that if m and n are natural numbers, then the complement of m in the sum $m + n$ is equivalent to n; the proof of this auxiliary fact is achieved by induction on n. Similar techniques prove that if E and F are finite sets, then so also are $E \times F$ and E^F, and, moreover, $\#(E \times F) = \#(E) \cdot \#(F)$ and $\#(E^F) = \#(E)^{\#(F)}$.

EXERCISE. The union of a finite set of finite sets is finite. If E is finite, then $\mathcal{P}(E)$ is finite and, moreover, $\#(\mathcal{P}(E)) = 2^{\#(E)}$. If E is a non-empty finite set of natural numbers, then there exists an element k in E such that $m \leq k$ for all m in E.

SECTION 14

ORDER

Throughout mathematics, and, in particular, for the generalization to infinite sets of the counting process appropriate to finite sets, the theory of order plays an important role. The basic definitions are simple. The only thing to remember is that the primary motivation comes from the familiar properties of "less than or equal to" and not "less than." There is no profound reason for this; it just happens that the generalization of "less than or equal to" occurs more frequently and is more amenable to algebraic treatment.

A relation R in a set X is called *antisymmetric* if, for every x and y in X, the simultaneous validity of $x\,R\,y$ and $y\,R\,x$ implies that $x = y$. A *partial order* (or sometimes simply an *order*) in a set X is a reflexive, antisymmetric, and transitive relation in X. It is customary to use only one symbol (or some typographically close relative of it) for most partial orders in most sets; the symbol in common use is the familiar inequality sign. Thus a partial order in X may be defined as a relation $\leq$ in X such that, for all x, y, and z in X, we have (i) $x \leq x$, (ii) if $x \leq y$ and $y \leq x$, then $x = y$, and (iii) if $x \leq y$ and $y \leq z$, then $x \leq z$. The reason for the qualifying "partial" is that some questions about order may be left unanswered. If for every x and y in X either $x \leq y$ or $y \leq x$, then $\leq$ is called a *total* (sometimes also *simple* or *linear*) order. A totally ordered set is frequently called a *chain*.

EXERCISE. Express the conditions of antisymmetry and totality for a relation R by means of equations involving R and its inverse.

The most natural example of a partial (and not total) order is inclusion. Explicitly: for each set X, the relation $\subset$ is a partial order in the power set $\mathcal{P}(X)$; it is a total order if and only if X is empty or X is a singleton. A well known example of a total order is the relation "less than or equal to" in the set of natural numbers. An interesting and frequently seen partial order is the relation of extension for functions. Explicitly: for given sets X and Y, let F be the set of all those functions whose domain is included in X and whose range is included in Y. Define a relation R in F by writing $f R g$ in case dom $f \subset$ dom g and $f(x) = g(x)$ for all x in dom f; in other words, $f R g$ means that f is a restriction of g, or, equivalently, that g is an extension of f. If we recall that the functions in F are, after all, certain subsets of the Cartesian product $X \times Y$, we recognize that $f R g$ means the same as $f \subset g$; extension is a special case of inclusion.

A *partially ordered set* is a set together with a partial order in it. A precise formulation of this "togetherness" goes as follows: a partially ordered set is an ordered pair $(X, \leqq)$, where X is a set and $\leqq$ is a partial order in X. This kind of definition is very common in mathematics; a mathematical structure is almost always a set "together" with some specified other sets, functions, and relations. The accepted way of making such definitions precise is by reference to ordered pairs, triples, or whatever is appropriate. That is not the only way. Observe, for instance, that knowledge of a partial order implies knowledge of its domain. If, therefore, we describe a partially ordered set as an ordered pair, we are being quite redundant; the second coordinate alone would have conveyed the same amount of information. In matters of language and notation, however, tradition always conquers pure reason. The accepted mathematical behavior (for structures in general, illustrated here for partially ordered sets) is to admit that ordered pairs are the right approach, to forget that the second coordinate is the important one, and to speak as if the first coordinate were all that mattered. Following custom, we shall often say something like "let X be a partially ordered set," when what we really mean is "let X be the domain of a partial order." The same linguistic conventions apply to totally ordered sets, i.e., to partially ordered sets whose order is in fact total.

The theory of partially ordered sets uses many words whose technical meaning is so near to their everyday connotation that they are almost self-explanatory. Suppose, to be specific, that X is a partially ordered set and that x and y are elements of X. We write $y \geqq x$ in case $x \leqq y$; in other words, $\geqq$ is the inverse of the relation $\leqq$. If $x \leqq y$ and $x \neq y$, we write $x < y$ and we say that x is *less* than or *smaller* than y, or that x is a *predecessor* of y. Alternatively, under the same circumstances, we write $y > x$ and we say that y is *greater* or *larger* than x, or y is a *successor* of x. The relation $<$ is such that (i) for no elements x and y do $x < y$ and $y < x$ hold simultaneously, and (ii) if $x < y$ and $y < z$, then $x < z$ (i.e., $<$ is transitive). If, conversely, $<$ is a relation in X satisfying (i) and (ii), and if $x \leqq y$ is defined to mean that either $x < y$ or $x = y$, then $\leqq$ is a partial order in X.

The connection between $\leqq$ and $<$ can be generalized to arbitrary relations. That is, given any relation R in a set X, we can define a relation S in X by writing $x \, S \, y$ in case $x \, R \, y$ but $x \neq y$, and, vice versa, given any relation S in X, we can define a relation R in X by writing $x \, R \, y$ in case either $x \, S \, y$ or $x = y$. To have an abbreviated way of referring to the passage from R to S and back, we shall say that S is the *strict* relation corresponding to R, and R is the *weak* relation corresponding to S. We shall say of a relation in a set X that it "partially orders X" in case either it is a partial order in X or else the corresponding weak relation is one.

If X is a partially ordered set, and if $a \, \epsilon \, X$, the set $\{x \, \epsilon \, X : x < a\}$ is the *initial segment* determined by a; we shall usually denote it by $s(a)$. The set $\{x \, \epsilon \, X : x \leqq a\}$ is the *weak initial segment* determined by a, and will be denoted by $\bar{s}(a)$. When it is important to emphasize the distinction between initial segments and weak initial segments, the former will be called *strict* initial segments. In general the words "strict" and "weak" refer to $<$ and $\leqq$ respectively. Thus, for instance, the initial segment determined by a may be described as the set of all predecessors of a, or, for emphasis, as the set of all *strict predecessors* of a; similarly the weak initial segment determined by a consists of all *weak predecessors* of a. If $x \leqq y$ and $y \leqq z$, we may say that y is *between* x and z; if $x < y$ and $y < z$, then y is *strictly between* x and z. If $x < y$ and if there is no element strictly between x and y, we say that x is an *immediate predecessor* of y, or y is an *immediate successor* of x.

If X is a partially ordered set (which may in particular be totally ordered), then it could happen that X has an element a such that $a \leqq x$ for every x in X. In that case we say that a is the *least* (*smallest, first*) element of X. The antisymmetry of an order implies that if X has a least element, then it has only one. If, similarly, X has an element a such that $x \leqq a$ for every x in X, then a is the *greatest* (*largest, last*) element of X; it too is unique (if it exists at all). The set ω of all natural numbers (with its customary ordering by magnitude) is an example of a partially ordered set with a first element (namely 0) but no last. The same set, but this time with the inverse ordering, has a last element but no first.

In partially ordered sets there is an important distinction between least elements and minimal ones. If, as before, X is a partially ordered set, an element a of X is called a *minimal* element of X in case there is no element in X strictly smaller than a. Equivalently, a is minimal if $x \leqq a$ implies that $x = a$. For an example, consider the collection $\mathcal{C}$ of non-empty subsets of a non-empty set X, with ordering by inclusion. Each singleton is a minimal element of $\mathcal{C}$, but clearly $\mathcal{C}$ has no least element (unless X itself is a singleton). We distinguish similarly between greatest and maximal elements; a *maximal* element of X is an element a such that X contains nothing strictly greater than a. Equivalently, a is maximal if $a \leq x$ implies that $x = a$.

An element a of a partially ordered set is said to be a *lower bound* of a subset E of X in case $a \leq x$ for every x in E; similarly a is an *upper bound* of E in case $x \leq a$ for every x in E. A set E may have no lower bounds or upper bounds at all, or it may have many; in the latter case it could happen that none of them belongs to E. (Examples?) Let E_* be the set of all lower bounds of E in X and let E^* be the set of all upper bounds of E in X. What was just said is that E_* may be empty, or $E_* \cap E$ may be empty. If $E_* \cap E$ is not empty, then it is a singleton consisting of the unique least element of E. Similar remarks apply, of course, to E^*. If it happens that the set E_* contains a greatest element a (necessarily unique), then a is called the *greatest lower bound* or *infimum* of E. The abbreviations *g.l.b.* and *inf* are in common use. Because of the difficulties in pronouncing the former, and even in remembering whether g.l.b. is up (greatest) or down (lower), we shall use the latter notation only. Thus inf E is the unique element in X (possibly not in E) that is a lower bound of E and that dominates (i.e., is greater than) every other lower bound of E. The definitions at the other end are completely parallel. If E^* has a least element a (necessarily unique), then a is called the *least upper bound* (*l.u.b.*) or *supremum* (*sup*) of E.

The ideas connected with partially ordered sets are easy to express but they take some time to assimilate. The reader is advised to manufacture many examples to illustrate the various possibilities in the behavior of partially ordered sets and their subsets. To aid him in this enterprise, we proceed to describe three special partially ordered sets with some amusing properties. (i) The set is $\omega \times \omega$. To avoid any possible confusion, we shall denote the order we are about to introduce by the neutral symbol R. If (a, b) and (x, y) are ordered pairs of natural numbers, then $(a, b) R (x, y)$ means, by definition, that $(2a + 1) \cdot 2^y \leq (2x + 1) \cdot 2^b$. (Here the inequality sign refers to the customary ordering of natural numbers.) The reader who is not willing to pretend ignorance of fractions will recognize that, except for notation, what we just defined is the usual order for $\dfrac{2a + 1}{2^b}$ and $\dfrac{2x + 1}{2^y}$. (ii) The set is $\omega \times \omega$ again. Once more we use a neutral symbol for the order; say S. If (a, b) and (x, y) are ordered pairs of natural numbers, then $(a, b)\ S\ (x, y)$ means, by definition, that either a is strictly less than x (in the customary sense), or else $a = x$ and $b \leq y$. Because of its resemblance to the way words are arranged in a dictionary, this is called the *lexicographical* order of $\omega \times \omega$. (iii) Once more the set is $\omega \times \omega$. The present order relation, say T, is such that $(a, b)\ T\ (x, y)$ means, by definition, that $a \leq x$ and $b \leq y$.

SECTION 15

THE AXIOM OF CHOICE

For the deepest results about partially ordered sets we need a new set-theoretic tool; we interrupt the development of the theory of order long enough to pick up that tool.

We begin by observing that a set is either empty or it is not, and, if it is not, then, by the definition of the empty set, there is an element in it. This remark can be generalized. If X and Y are sets, and if one of them is empty, then the Cartesian product $X \times Y$ is empty. If neither X nor Y is empty, then there is an element x in X, and there is an element y in Y; it follows that the ordered pair (x, y) belongs to the Cartesian product $X \times Y$, so that $X \times Y$ is not empty. The preceding remarks constitute the cases $n = 1$ and $n = 2$ of the following assertion: if $\{X_i\}$ is a finite sequence of sets, for i in n, say, then a necessary and sufficient condition that their Cartesian product be empty is that at least one of them be empty. The assertion is easy to prove by induction on n. (The case $n = 0$ leads to a slippery argument about the empty function; the uninterested reader may start his induction at 1 instead of 0.)

The generalization to infinite families of the non-trivial part of the assertion in the preceding paragraph (necessity) is the following important principle of set theory.

Axiom of choice. *The Cartesian product of a non-empty family of non-empty sets is non-empty.*

In other words: if $\{X_i\}$ is a family of non-empty sets indexed by a non-empty set I, then there exists a family $\{x_i\}$, $i \in I$, such that $x_i \in X_i$ for each i in I.

Suppose that $\mathcal{C}$ is a non-empty collection of non-empty sets. We may regard $\mathcal{C}$ as a family, or, to say it better, we can convert $\mathcal{C}$ into an indexed set, just by using the collection $\mathcal{C}$ itself in the role of the index set and using the identity mapping on $\mathcal{C}$ in the role of the indexing. The axiom of choice then says that the Cartesian product of the sets of $\mathcal{C}$ has at least one element. An element of such a Cartesian product is, by definition, a function (family, indexed set) whose domain is the index set (in this case $\mathcal{C}$) and whose value at each index belongs to the set bearing that index. Conclusion: there exists a function f with domain $\mathcal{C}$ such that if $A \in \mathcal{C}$, then $f(A) \in A$. This conclusion applies, in particular, in case $\mathcal{C}$ is the collection of all non-empty subsets of a non-empty set X. The assertion in that case is that there exists a function f with domain $\mathcal{P}(X) - \{\varnothing\}$ such that if A is in that domain, then $f(A) \in A$. In intuitive language the function f can be described as a simultaneous choice of an element from each of many sets; this is the reason for the name of the axiom. (A function that in this sense "chooses" an element out of each non-empty subset of a set X is called a *choice function* for X.) We have seen that if the collection of sets we are choosing from is finite, then the possibility of simultaneous choice is an easy consequence of what we knew before the axiom of choice was even stated; the role of the axiom is to guarantee that possibility in infinite cases.

The two consequences of the axiom of choice in the preceding paragraph (one for the power set of a set and the other for more general collections of sets) are in fact just reformulations of that axiom. It used to be considered important to examine, for each consequence of the axiom of choice, the extent to which the axiom is needed in the proof of the consequence. An alternative proof without the axiom of choice spelled victory; a converse proof, showing that the consequence is equivalent to the axiom of choice (in the presence of the remaining axioms of set theory) meant honorable defeat. Anything in between was considered exasperating. As a sample (and an exercise) we mention the assertion that every relation includes a function with the same domain. Another sample: if $\mathcal{C}$ is a collection of pairwise disjoint non-empty sets, then there exists a set A such that $A \cap C$ is a singleton for each C in $\mathcal{C}$. Both these assertions are among the many known to be equivalent to the axiom of choice.

As an illustration of the use of the axiom of choice, consider the assertion that if a set is infinite, then it has a subset equivalent to ω. An informal argument might run as follows. If X is infinite, then, in particular, it is not empty (that is, it is not equivalent to 0); hence it has an element, say x_0. Since X is not equivalent to 1, the set $X - \{x_0\}$ is not empty; hence it has an element, say x_1. Repeat this argument ad infinitum; the next step, for instance, is to say that $X - \{x_0, x_1\}$ is not empty, and, therefore, it has an element, say x_2. The result is an infinite sequence $\{x_n\}$ of distinct elements of X; q.e.d. This sketch of a proof at least has the virtue of being honest about the most important idea behind it; the act of choosing an element from a non-empty set was repeated infinitely often. The mathematician experienced in the ways of the axiom of choice will often offer such an informal argument; his experience enables him to see at a glance how to make it precise. For our purposes it is advisable to take a longer look.

Let f be a choice function for X; that is, f is a function from the collection of all non-empty subsets of X to X such that $f(A) \in A$ for all A in the domain of f. Let $\mathcal{C}$ be the collection of all finite subsets of X. Since X is infinite, it follows that if $A \in \mathcal{C}$, then $X - A$ is not empty, and hence that $X - A$ belongs to the domain of f. Define a function g from $\mathcal{C}$ to $\mathcal{C}$ by writing $g(A) = A \cup \{f(X - A)\}$. In words: $g(A)$ is obtained by adjoining to A the element that f chooses from $X - A$. We apply the recursion theorem to the function g; we may start it rolling with, for instance, the set $\varnothing$. The result is that there exists a function U from ω into $\mathcal{C}$ such that $U(0) = \varnothing$ and $U(n^+) = U(n) \cup \{f(X - U(n))\}$ for every natural number n. Assertion: if $v(n) = f(X - U(n))$, then v is a one-to-one correspondence from ω to X, and hence, indeed, ω is equivalent to some subset of X (namely the range of v). To prove the assertion, we make a series of elementary observations; their proofs are easy consequences of the definitions. First: $v(n) \in' U(n)$ for all n. Second: $v(n) \in U(n^+)$ for all n. Third: if n and m are natural numbers and $n \leqq m$, then $U(n) \subset U(m)$. Fourth: if n and m are natural numbers and $n < m$, then $v(n) \neq v(m)$. (Reason: $v(n) \in U(m)$ but $v(m) \in' U(m)$.) The last observation implies that v maps distinct natural numbers onto distinct elements of X; all we have to remember is that of any two distinct natural numbers one of them is strictly smaller than the other.

The proof is complete; we know now that every infinite set has a subset equivalent to ω. This result, proved here not so much for its intrinsic interest as for an example of the proper use of the axiom of choice, has an interesting corollary. The assertion is that a set is infinite if and only if it is equivalent to a proper subset of itself. The "if" we already know; it says merely that a finite set cannot be equivalent to a proper subset. To prove the "only if," suppose that X is infinite, and let v be a one-to-one correspondence from ω into X. If x is in the range of v, say $x = v(n)$, write $h(x) = v(n^+)$; if x is not in the range of v, write $h(x) = x$. It is easy to verify that h is a one-to-one correspondence from X into itself. Since the range of h is a proper subset of X (it does not contain $v(0)$), the proof of the corollary is complete. The assertion of the corollary was used by Dedekind as the very definition of infinity.

SECTION 16

ZORN'S LEMMA

An existence theorem asserts the existence of an object belonging to a certain set and possessing certain properties. Many existence theorems can be formulated (or, if need be, reformulated) so that the underlying set is a partially ordered set and the crucial property is maximality. Our next purpose is to state and prove the most important theorem of this kind.

Zorn's lemma. *If X is a partially ordered set such that every chain in X has an upper bound, then X contains a maximal element.*

DISCUSSION. Recall that a chain is a totally ordered set. By a chain "in X" we mean a subset of X such that the subset, considered as a partially ordered set on its own right, turns out to be totally ordered. If A is a chain in X, the hypothesis of Zorn's lemma guarantees the existence of an upper bound for A in X; it does not guarantee the existence of an upper bound for A in A. The conclusion of Zorn's lemma is the existence of an element a in X with the property that if $a \leqq x$, then necessarily $a = x$.

The basic idea of the proof is similar to the one used in our preceding discussion of infinite sets. Since, by hypothesis, X is not empty, it has an element, say x_0. If x_0 is maximal, stop here. If it is not, then there exists an element, say x_1, strictly greater than x_0. If x_1 is maximal, stop here; otherwise continue. Repeat this argument ad infinitum; ultimately it must lead to a maximal element.

The last sentence is probably the least convincing part of the argument; it hides a multitude of difficulties. Observe, for instance, the following possibility. It could happen that the argument, repeated ad infinitum, leads to a whole infinite sequence of non-maximal elements; what are we to do in that case? The answer is that the range of such an infinite sequence is a chain in X, and, consequently, has an upper bound; the thing to do is to start the whole argument all over again, beginning with that

62

upper bound. Just exactly when and how all this comes to an end is obscure, to say the least. There is no help for it; we must look at the precise proof. The structure of the proof is an adaptation of one originally given by Zermelo.

PROOF. The first step is to replace the abstract partial ordering by the inclusion order in a suitable collection of sets. More precisely, we consider, for each element x in X, the weak initial segment $\bar{s}(x)$ consisting of x and all its predecessors. The range S of the function $\bar{s}$ (from X to $\mathcal{P}(X)$) is a certain collection of subsets of X, which we may, of course, regard as (partially) ordered by inclusion. The function $\bar{s}$ is one-to-one, and a necessary and sufficient condition that $\bar{s}(x) \subset \bar{s}(y)$ is that $x \leqq y$. In view of this, the task of finding a maximal element in X is the same as the task of finding a maximal set in S. The hypothesis about chains in X implies (and is, in fact, equivalent to) the corresponding statement about chains in S.

Let $\mathfrak{X}$ be the set of all chains in X; every member of $\mathfrak{X}$ is included in $\bar{s}(x)$ for some x in X. The collection $\mathfrak{X}$ is a non-empty collection of sets, partially ordered by inclusion, and such that if $\mathcal{C}$ is a chain in $\mathfrak{X}$, then the union of the sets in $\mathcal{C}$ (i.e., $\bigcup_{A \,\epsilon\, \mathcal{C}} A$) belongs to $\mathfrak{X}$. Since each set in $\mathfrak{X}$ is dominated by some set in S, the passage from S to $\mathfrak{X}$ cannot introduce any new maximal elements. One advantage of the collection $\mathfrak{X}$ is the slightly more specific form that the chain hypothesis assumes; instead of saying that each chain $\mathcal{C}$ has some upper bound in S, we can say explicitly that the union of the sets of $\mathcal{C}$, which is clearly an upper bound of $\mathcal{C}$, is an element of the collection $\mathfrak{X}$. Another technical advantage of $\mathfrak{X}$ is that it contains all the subsets of each of its sets; this makes it possible to enlarge non-maximal sets in $\mathfrak{X}$ slowly, one element at a time.

Now we can forget about the given partial order in X. In what follows we consider a non-empty collection $\mathfrak{X}$ of subsets of a non-empty set X, subject to two conditions: every subset of each set in $\mathfrak{X}$ is in $\mathfrak{X}$, and the union of each chain of sets in $\mathfrak{X}$ is in $\mathfrak{X}$. Note that the first condition implies that $\varnothing \,\epsilon\, \mathfrak{X}$. Our task is to prove that there exists in $\mathfrak{X}$ a maximal set.

Let f be a choice function for X, that is, f is a function from the collection of all non-empty subsets of X to X such that $f(A) \,\epsilon\, A$ for all A in the domain of f. For each set A in $\mathfrak{X}$, let $\hat{A}$ be the set of all those elements x of X whose adjunction to A produces a set in $\mathfrak{X}$; in other words, $\hat{A} = \{x \,\epsilon\, X \colon A \cup \{x\} \,\epsilon\, \mathfrak{X}\}$. Define a function g from $\mathfrak{X}$ to $\mathfrak{X}$ as follows: if $\hat{A} - A \neq \varnothing$, then $g(A) = A \cup \{f(\hat{A} - A)\}$; if $\hat{A} - A = \varnothing$, then $g(A) = A$. It follows from the definition of $\hat{A}$ that $\hat{A} - A = \varnothing$ if and only if A is maximal. In these terms, therefore, what we must prove is that there exists in $\mathfrak{X}$ a set A such that $g(A) = A$. It turns out that the crucial property of g is the fact that $g(A)$ (which always includes A) contains at most one more element than A.

Now, to facilitate the exposition, we introduce a temporary definition. We shall say that a subcollection $\mathfrak{J}$ of $\mathfrak{X}$ is a *tower* if

(i) $\varnothing \in \mathfrak{J}$,

(ii) *if $A \in \mathfrak{J}$, then $g(A) \in \mathfrak{J}$,*

(iii) *if $\mathfrak{C}$ is a chain in $\mathfrak{J}$, then $\bigcup_{A \in \mathfrak{C}} A \in \mathfrak{J}$.*

Towers surely exist; the whole collection $\mathfrak{X}$ is one. Since the intersection of a collection of towers is again a tower, it follows, in particular, that if $\mathfrak{J}_0$ is the intersection of all towers, then $\mathfrak{J}_0$ is the smallest tower. Our immediate purpose is to prove that the tower $\mathfrak{J}_0$ is a chain.

Let us say that a set C in $\mathfrak{J}_0$ is *comparable* if it is comparable with every set in $\mathfrak{J}_0$; this means that if $A \in \mathfrak{J}_0$, then either $A \subset C$ or $C \subset A$. To say that $\mathfrak{J}_0$ is a chain means that all the sets in $\mathfrak{J}_0$ are comparable. Comparable sets surely exist; $\varnothing$ is one of them. In the next couple of paragraphs we concentrate our attention on an arbitrary but temporarily fixed comparable set C.

Suppose that $A \in \mathfrak{J}_0$ and A is a proper subset of C. Assertion: $g(A) \subset C$. The reason is that since C is comparable, either $g(A) \subset C$ or C is a proper subset of $g(A)$. In the latter case A is a proper subset of a proper subset of $g(A)$, and this contradicts the fact that $g(A) - A$ cannot be more than a singleton.

Consider next the collection $\mathfrak{U}$ of all those sets A in $\mathfrak{J}_0$ for which either $A \subset C$ or $g(C) \subset A$. The collection $\mathfrak{U}$ is somewhat smaller than the collection of sets in $\mathfrak{J}_0$ comparable with $g(C)$; indeed if $A \in \mathfrak{U}$, then, since $C \subset g(C)$, either $A \subset g(C)$ or $g(C) \subset A$. Assertion: $\mathfrak{U}$ is a tower. Since $\varnothing \subset C$, the first condition on towers is satisfied. To prove the second condition, i.e., that if $A \in \mathfrak{U}$, then $g(A) \in \mathfrak{U}$, split the discussion into three cases. First: A is a proper subset of C. Then $g(A) \subset C$ by the preceding paragraph, and therefore $g(A) \in \mathfrak{U}$. Second: $A = C$. Then $g(A) = g(C)$, so that $g(C) \subset g(A)$, and therefore $g(A) \in \mathfrak{U}$. Third: $g(C) \subset A$. Then $g(C) \subset g(A)$, and therefore $g(A) \in \mathfrak{U}$. The third condition on towers, i.e., that the union of a chain in $\mathfrak{U}$ belongs to $\mathfrak{U}$, is immediate from the definition of $\mathfrak{U}$. Conclusion: $\mathfrak{U}$ is a tower included in $\mathfrak{J}_0$, and therefore, since $\mathfrak{J}_0$ is the smallest tower, $\mathfrak{U} = \mathfrak{J}_0$.

The preceding considerations imply that for each comparable set C the set $g(C)$ is comparable also. Reason: given C, form $\mathfrak{U}$ as above; the fact that $\mathfrak{U} = \mathfrak{J}_0$ means that if $A \in \mathfrak{J}_0$, then either $A \subset C$ (in which case $A \subset g(C)$) or $g(C) \subset A$.

We now know that $\varnothing$ is comparable and that g maps comparable sets onto comparable sets. Since the union of a chain of comparable sets is comparable, it follows that the comparable sets (in $\mathfrak{I}_0$) constitute a tower, and hence that they exhaust $\mathfrak{I}_0$; this is what we set out to prove about $\mathfrak{I}_0$.

Since $\mathfrak{I}_0$ is a chain, the union, say A, of all the sets in $\mathfrak{I}_0$ is itself a set in $\mathfrak{I}_0$. Since the union includes all the sets in $\mathfrak{I}_0$, it follows that $g(A) \subset A$. Since always $A \subset g(A)$, it follows that $A = g(A)$, and the proof of Zorn's lemma is complete.

EXERCISE. Zorn's lemma is equivalent to the axiom of choice. [Hint for the proof: given a set X, consider functions f such that $\operatorname{dom} f \subset \mathcal{P}(X)$, $\operatorname{ran} f \subset X$, and $f(A) \in A$ for all A in $\operatorname{dom} f$; order these functions by extension, use Zorn's lemma to find a maximal one among them, and prove that if f is maximal, then $\operatorname{dom} f = \mathcal{P}(X) - \{\varnothing\}$.] Consider each of the following statements and prove that they too are equivalent to the axiom of choice. (i) Every partially ordered set has a maximal chain (i.e., a chain that is not a proper subset of any other chain). (ii) Every chain in a partially ordered set is included in some maximal chain. (iii) Every partially ordered set in which each chain has a least upper bound has a maximal element.

SECTION 17

WELL ORDERING

A partially ordered set may not have a smallest element, and, even if it has one, it is perfectly possible that some subset will fail to have one. A partially ordered set is called *well ordered* (and its ordering is called a *well ordering*) if every non-empty subset of it has a smallest element. One consequence of this definition, worth noting even before we look at any examples and counterexamples, is that every well ordered set is totally ordered. Indeed, if x and y are elements of a well ordered set, then $\{x, y\}$ is a non-empty subset of that well ordered set and has therefore a first element; according as that first element is x or y, we have $x \leqq y$ or $y \leqq x$.

For each natural number n, the set of all predecessors of n (that is, in accordance with our definitions, the set n) is a well ordered set (ordered by magnitude), and the same is true of the set ω of all natural numbers. The set $\omega \times \omega$, with $(a, b) \leqq (x, y)$ defined to mean $(2a + 1)2^y \leqq (2x + 1)2^b$ is not well ordered. One way to see this is to note that $(a, b + 1) \leqq (a, b)$ for all a and b; it follows that the entire set $\omega \times \omega$ has no least element. Some subsets of $\omega \times \omega$ do have a least element. Consider, for example, the set E of all those pairs (a, b) for which $(1, 1) \leqq (a, b)$; the set E has $(1, 1)$ for its least element. Caution: E, considered as a partially ordered set on its own right, is still not well ordered. The trouble is that even though E has a least element, many subsets of E fail to have one; for an example consider the set of all those pairs (a, b) in E for which $(a, b) \neq (1, 1)$. One more example: $\omega \times \omega$ is well ordered by its lexicographical ordering.

One of the pleasantest facts about well ordered sets is that we can prove things about their elements by a process similar to mathematical induction. Precisely speaking, suppose that S is a subset of a well ordered set X, and suppose that whenever an element x of X is such that the entire initial segment $s(x)$ is included in S, then x itself belongs to S; the **principle of transfinite induction** asserts that under these circumstances we must

66

have $S = X$. Equivalently: if the presence in a set of all the strict predecessors of an element always implies the presence of the element itself,
then the set must contain everything.

A few remarks are in order before we look at the proof. The statement
of the ordinary principle of mathematical induction differs from that of
transfinite induction in two conspicuous respects. One: the latter, instead
of passing to each element from its predecessor, passes to each element
from the set of all its predecessors. Two: in the latter there is no assumption about a starting element (such as zero). The first difference is important: an element in a well ordered set may fail to have an immediate predecessor. The present statement when applied to ω is easily proved to be
equivalent to the principle of mathematical induction; that principle,
however, when applied to an arbitrary well ordered set, is not equivalent
to the principle of transfinite induction. To put it differently: the two
statements are in general not equivalent to each other; their equivalence
in ω is a happy but special circumstance.

Here is an example. Let X be ω^+, i.e., $X = \omega \cup \{\omega\}$. Define order in
X by ordering the elements of ω as usual and by requiring that $n < \omega$ for
all n in ω. The result is a well ordered set. Question: does there exist a
proper subset S of X such that $0 \epsilon S$ and such that $n + 1 \epsilon S$ whenever
$n \epsilon S$? Answer: yes, namely $S = \omega$.

The second difference between ordinary induction and transfinite induction (no starting element required for the latter) is more linguistic than
conceptual. If x_0 is the smallest element of X, then $s(x_0)$ is empty, and,
consequently, $s(x_0) \subset S$; the hypothesis of the principle of transfinite induction requires therefore that x_0 belong to S.

The proof of the principle of transfinite induction is almost trivial. If
$X - S$ is not empty, then it has a smallest element, say x. This implies
that every element of the initial segment $s(x)$ belongs to S, and hence, by
the induction hypothesis, that x belongs to S. This is a contradiction
(x cannot belong to both S and $X - S$); the conclusion is that $X - S$ is
empty after all.

We shall say that a well ordered set A is a *continuation* of a well ordered
set B, if, in the first place, B is a subset of A, if, in fact, B is an initial segment of A, and if, finally, the ordering of the elements in B is the same as
their ordering in A. Thus if X is a well ordered set and if a and b are elements of X with $b < a$, then $s(a)$ is a continuation of $s(b)$, and, of course,
X is a continuation of both $s(a)$ and $s(b)$.

If $\mathcal{C}$ is an arbitrary collection of initial segments of a well ordered set,
then $\mathcal{C}$ is a chain with respect to continuation; this means that $\mathcal{C}$ is a collec-

tion of well ordered sets with the property that of any two distinct members of the collection one is a continuation of the other. A sort of converse of this comment is also true and is frequently useful. If a collection $\mathfrak{C}$ of well ordered sets is a chain with respect to continuation, and if U is the union of the sets of $\mathfrak{C}$, then there is a unique well ordering of U such that U is a continuation of each set (distinct from U itself) in the collection $\mathfrak{C}$. Roughly speaking, the union of a chain of well ordered sets is well ordered. This abbreviated formulation is dangerous because it does not explain that "chain" is meant with respect to continuation. If the ordering implied by the word "chain" is taken to be simply order-preserving inclusion, then the conclusion is not valid.

The proof is straightforward. If a and b are in U, then there exist sets A and B in $\mathfrak{C}$ with $a \in A$ and $b \in B$. Since either $A = B$ or one of A and B is a continuation of the other, it follows that in every case both a and b belong to some one set in $\mathfrak{C}$; the order of U is defined by ordering each pair $\{a, b\}$ the way it is ordered in any set of $\mathfrak{C}$ that contains both a and b. Since $\mathfrak{C}$ is a chain, this order is unambiguously determined. (An alternative way of defining the promised order in U is to recall that the given orders, in the sets of $\mathfrak{C}$, are sets of ordered pairs, and to form the union of all those sets of ordered pairs.)

A direct verification shows that the relation defined in the preceding paragraph is indeed an order, and that, moreover, its construction was forced on us at every step (i.e., that the final order is uniquely determined by the given orders). The proof that the result is actually a well ordering is equally direct. Each non-empty subset of U must have a non-empty intersection with some set in $\mathfrak{C}$, and hence it must have a first element in that set; the fact that $\mathfrak{C}$ is a continuation chain implies that that first element is necessarily the first element of U also.

EXERCISE. A subset A of a partially ordered set X is *cofinal* in X in case for each element x of X there exists an element a of A such that $x \leq a$. Prove that every totally ordered set has a cofinal well ordered subset.

The importance of well ordering stems from the following result, from which we may infer, among other things, that the principle of transfinite induction is much more widely applicable than a casual glance might indicate.

Well ordering theorem. *Every set can be well ordered.*

DISCUSSION. A better (but less traditional) statement is this: for each set X, there is a well ordering with domain X. Warning: the well ordering is not promised to have any relation whatsoever to any other structure that the given set might already possess. If, for instance, the reader knows of some partially or totally ordered sets whose ordering is very definitely not a well ordering, he should not jump to the conclusion that he has discovered a paradox. The only conclusion to be drawn is that some sets can be ordered in many ways, some of which are well orderings and others are not, and we already knew that.

PROOF. We apply Zorn's lemma. Given the set X, consider the collection $\mathcal{W}$ of all well ordered subsets of X. Explicitly: an element of $\mathcal{W}$ is a subset A of X together with a well ordering of A. We partially order $\mathcal{W}$ by continuation.

The collection $\mathcal{W}$ is not empty, because, for instance, $\varnothing \in \mathcal{W}$. If $X \neq \varnothing$, less annoying elements of $\mathcal{W}$ can be exhibited; one such is $\{(x, x)\}$, for any particular element x of X. If $\mathcal{C}$ is a chain in $\mathcal{W}$, then the union U of the sets in $\mathcal{C}$ has a unique well ordering that makes U "larger" than (or equal to) each set in $\mathcal{C}$; this is exactly what our preceding discussion of continuation has accomplished. This means that the principal hypothesis of Zorn's lemma has been verified; the conclusion is that there exists a maximal well ordered set, say M, in $\mathcal{W}$. The set M must be equal to the entire set X. Reason: if x is an element of X not in M, then M can be enlarged by putting x after all the elements of M. The rigorous formulation of this unambiguous but informal description is left as an exercise for the reader. With that out of the way, the proof of the well ordering theorem is complete.

EXERCISE. Prove that a totally ordered set is well ordered if and only if the set of strict predecessors of each element is well ordered. Does any such condition apply to partially ordered sets? Prove that the well ordering theorem implies the axiom of choice (and hence is equivalent to that axiom and to Zorn's lemma). Prove that if R is a partial order in a set X, then there exists a total order S in X such that $R \subset S$; in other words, every partial order can be extended to a total order without enlarging the domain.

SECTION 18

TRANSFINITE RECURSION

The process of "definition by induction" has a transfinite analogue. The ordinary recursion theorem constructs a function on ω; the raw material is a way of getting the value of the function at each non-zero element n of ω from its value at the element preceding n. The transfinite analogue constructs a function on any well ordered set W; the raw material is a way of getting the value of the function at each element a of W from its values at all the predecessors of a.

To be able to state the result concisely, we introduce some auxiliary concepts. If a is an element of a well ordered set W, and if X is an arbitrary set, then by a *sequence of type a in X* we shall mean a function from the initial segment of a in W into X. The sequences of type a, for a in ω^+, are just what we called sequences before, finite or infinite according as $a < \omega$ or $a = \omega$. If U is a function from W to X, then the restriction of U to the initial segment $s(a)$ of a is an example of a sequence of type a for each a in W; in what follows we shall find it convenient to denote that sequence by U^a (instead of $U \mid s(a)$).

A *sequence function of type W in X* is a function f whose domain consists of all sequences of type a in X, for all elements a in W, and whose range is included in X. Roughly speaking, a sequence function tells us how to "lengthen" a sequence; given a sequence that stretches up to (but not including) some element of W we can use a sequence function to tack on one more term.

Transfinite recursion theorem. *If W is a well ordered set, and if f is a sequence function of type W in a set X, then there exists a unique function U from W into X such that $U(a) = f(U^a)$ for each a in W.*

70

PROOF. The proof of uniqueness is an easy transfinite induction. To prove existence, recall that a function from W to X is a certain kind of subset of $W \times X$; we shall construct U explicitly as a set of ordered pairs. Call a subset A of $W \times X$ *f-closed* if it has the following property: whenever $a \in W$ and t is a sequence of type a included in A (that is, $(c, t(c)) \in A$ for all c in the initial segment $s(a)$), then $(a, f(t)) \in A$. Since $W \times X$ itself is f-closed, such sets do exist; let U be the intersection of them all. Since U itself is f-closed, it remains only to prove that U is a function. We are to prove, in other words, that for each c in W there exists at most one element x in X such that $(c, x) \in U$. (Explicitly: if both (c, x) and (c, y) belong to U, then $x = y$.) The proof is inductive. Let S be the set of all those elements c of W for which it is indeed true that $(c, x) \in U$ for at most one x. We shall prove that if $s(a) \subset S$, then $a \in S$.

To say that $s(a) \subset S$ means that if $c < a$ in W, then there exists a unique element x in X such that $(c, x) \in U$. The correspondence $c \to x$ thereby defined is a sequence of type a, say t, and $t \subset U$. If a does not belong to S, then $(a, y) \in U$ for some y different from $f(t)$. Assertion: the set $U - \{(a, y)\}$ is f-closed. This means that if $b \in W$ and if r is a sequence of type b included in $U - \{(a, y)\}$, then $(b, f(r)) \in U - \{(a, y)\}$. Indeed, if $b = a$, then r must be t (by the uniqueness assertion of the theorem), and the reason the diminished set contains $(b, f(r))$ is that $f(t) \neq y$; if, on the other hand, $b \neq a$, then the reason the diminished set contains $(b, f(r))$ is that U is f-closed (and $b \neq a$). This contradicts the fact that U is the smallest f-closed set, and we may conclude that $a \in S$.

The proof of the existence assertion of the transfinite recursion theorem is complete. An application of the transfinite recursion theorem is called *definition by transfinite induction*.

We continue with an important part of the theory of order, which, incidentally, will also serve as an illustration of how the transfinite recursion theorem can be applied.

Two partially ordered sets (which may in particular be totally ordered and even well ordered) are called *similar* if there exists an order-preserving one-to-one correspondence between them. More explicitly: to say of the partially ordered sets X and Y that they are similar (in symbols $X \cong Y$) means that there exists a one-to-one correspondence, say f, from X onto Y, such that if a and b are in X, then a necessary and sufficient condition that $f(a) \leq f(b)$ (in Y) is that $a \leq b$ (in X). A correspondence such as f is sometimes called a *similarity*.

EXERCISE. Prove that a similarity preserves $<$ (in the same sense in which the definition demands the preservation of $\leq$) and that, in fact, a one-to-one function that maps one partially ordered set onto another is a similarity if and only if it preserves $<$.

The identity mapping on a partially ordered set X is a similarity from X onto X. If X and Y are partially ordered sets and if f is a similarity from X onto Y, then (since f is one-to-one) there exists an unambiguously determined inverse function f^{-1} from Y onto X, and f^{-1} is a similarity. If, moreover, g is a similarity from Y onto a partially ordered set Z, then the composite gf is a similarity from X onto Z. It follows from these comments that if we restrict attention to some particular set E, and if, accordingly, we consider only such partial orders whose domain is a subset of E, then similarity is an equivalence relation in the set of partially ordered sets so obtained. The same is true if we narrow the field even further and consider only well orderings whose domain is included in E; similarity is an equivalence relation in the set of well ordered sets so obtained. Although similarity was defined for partially ordered sets in complete generality, and the subject can be studied on that level, our interest in what follows will be in similarity for well ordered sets only.

It is easily possible for a well ordered set to be similar to a proper subset; for an example consider the set of all natural numbers and the set of all even numbers. (As always, a natural number m is defined to be even if there exists a natural number n such that $m = 2n$. The mapping $n \rightarrow 2n$ is a similarity from the set of all natural numbers onto the set of all even numbers.) A similarity of a well ordered set with a part of itself is, however, a very special kind of mapping. If, in fact, f is a similarity of a well ordered set X into itself, then $a \leqq f(a)$ for each a in X. The proof is based directly on the definition of well ordering. If there are elements b such that $f(b) < b$, then there is a least one among them. If $a < b$, where b is that least one, then $a \leqq f(a)$; it follows, in particular, with $a = f(b)$, that $f(b) \leqq f(f(b))$. Since, however, $f(b) < b$, the order-preserving character of f implies that $f(f(b)) < f(b)$. The only way out of the contradiction is to admit the impossibility of $f(b) < b$.

The result of the preceding paragraph has three especially useful consequences. The first of these is the fact that if two well ordered sets, X and Y say, are similar at all, then there is just one similarity between them. Suppose indeed that both g and h are similarities from X onto Y, and write $f = g^{-1}h$. Since f is a similarity of X onto itself, it follows that $a \leqq f(a)$ for each a in X. This means that $a \leqq g^{-1}(h(a))$ for each a in X. Applying g, we infer that $g(a) \leqq h(a)$ for each a in X. The situation is symmetric in g and h, so that we may also infer that $h(a) \leqq g(a)$ for each a in X. Conclusion: $g = h$.

A second consequence is the fact that a well ordered set is never similar to one of its initial segments. If, indeed, X is a well ordered set, a is an element of X, and f is a similarity from X onto $s(a)$, then, in particular, $f(a) \in s(a)$, so that $f(a) < a$, and that is impossible.

The third and chief consequence is the comparability theorem for well ordered sets. The assertion is that if X and Y are well ordered sets, then either X and Y are similar, or one of them is similar to an initial segment of the other. Just for practice we shall use the transfinite recursion theorem in the proof, although it is perfectly easy to avoid it. We assume that X and Y are non-empty well ordered sets such that neither is similar to an initial segment of the other; we proceed to prove that under these circumstances X must be similar to Y. Suppose that $a \in X$ and that t is a sequence of type a in Y; in other words t is a function from $s(a)$ into Y. Let $f(t)$ be the least of the proper upper bounds of the range of t in Y, if there are any; in the contrary case, let $f(t)$ be the least element of Y. In the terminology of the transfinite recursion theorem, the function f thereby determined is a sequence function of type X in Y. Let U be the function that the transfinite recursion theorem associates with this situation. An easy argument (by transfinite induction) shows that, for each z in X, the function U maps the initial segment determined by a in X one-to-one onto the initial segment determined by $U(a)$ in Y. This implies that U is a similarity, and the proof is complete.

Here is a sketch of an alternative proof that does not use the transfinite recursion theorem. Let X_0 be the set of those elements a of X for which there exists an element b of Y such that $s(a)$ is similar to $s(b)$. For each a in X_0, write $U(a)$ for the corresponding (uniquely determined) b in Y, and let Y_0 be the range of U. It follows that either $X_0 = X$, or else X_0 is an initial segment of X and $Y_0 = Y$.

EXERCISE. Each subset of a well ordered set X is similar either to X or to an initial segment of X. If X and Y are well ordered sets and $X \cong Y$ (i.e., X is similar to Y), then the similarity maps the least upper bound (if any) of each subset of X onto the least upper bound of the image of that subset.

SECTION 19

ORDINAL NUMBERS

The successor x^+ of a set x was defined as $x \cup \{x\}$, and then ω was constructed as the smallest set that contains 0 and that contains x^+ whenever it contains x. What happens if we start with ω, form its successor ω^+, then form the successor of that, and proceed so on ad infinitum? In other words: is there something out beyond ω, ω^+, $(\omega^+)^+$, $\cdots$, etc., in the same sense in which ω is beyond 0, 1, 2, $\cdots$, etc.?

The question calls for a set, say T, containing ω, such that each element of T (other than ω itself) can be obtained from ω by the repeated formation of successors. To formulate this requirement more precisely we introduce some special and temporary terminology. Let us say that a function f whose domain is the set of strict predecessors of some natural number n (in other words, dom $f = n$) is an ω-*successor function* if $f(0) = \omega$ (provided that $n \neq 0$, so that $0 < n$), and $f(m^+) = (f(m))^+$ whenever $m^+ < n$. An easy proof by mathematical induction shows that for each natural number n there exists a unique ω-successor function with domain n. To say that something is either equal to ω or can be obtained from ω by the repeated formation of successors means that it belongs to the range of some ω-successor function. Let $S(n, x)$ be the sentence that says "n is a natural number and x belongs to the range of the ω-successor function with domain n." A set T such that $x \in T$ if and only if $S(n, x)$ is true for some n is what we are looking for; such a set is as far beyond ω as ω is beyond 0.

We know that for each natural number n we are permitted to form the set $\{x : S(n, x)\}$. In other words, for each natural number n, there exists a set $F(n)$ such that $x \in F(n)$ if and only if $S(n, x)$ is true. The connection between n and $F(n)$ looks very much like a function. It turns out, how-

ever, that none of the methods of set construction that we have seen so
far is sufficiently strong to prove the existence of a set F of ordered pairs
such that $(n, x) \in F$ if and only if $x \in F(n)$. To achieve this obviously desirable state of affairs, we need one more set-theoretic principle (our last).
The new principle says, roughly speaking, that anything intelligent that
one can do to the elements of a set yields a set.

Axiom of substitution. *If $S(a, b)$ is a sentence such that for each a in a
set A the set $\{b : S(a, b)\}$ can be formed, then there exists a function F with
domain A such that $F(a) = \{b : S(a, b)\}$ for each a in A.*

To say that $\{b : S(a, b)\}$ can be formed means, of course, that there exists
a set $F(a)$ such that $b \in F(a)$ if and only if $S(a, b)$ is true. The axiom of
extension implies that the function described in the axiom of substitution
is uniquely determined by the given sentence and the given set. The reason for the name of the axiom is that it enables us to make a new set out
of an old one by substituting something new for each element of the old.

The chief application of the axiom of substitution is in extending the
process of counting far beyond the natural numbers. From the present
point of view, the crucial property of a natural number is that it is a well
ordered set such that the initial segment determined by each element is
equal to that element. (Recall that if m and n are natural numbers, then
$m < n$ means $m \in n$; this implies that $\{m \in \omega : m < n\} = n$.) This is the
property on which the extended counting process is based; the fundamen-
tal definition in this circle of ideas is due to von Neumann. An *ordinal
number* is defined as a well ordered set α such that $s(\xi) = \xi$ for all ξ in α;
here $s(\xi)$ is, as before, the initial segment $\{\eta \in \alpha : \eta < \xi\}$.

An example of an ordinal number that is not a natural number is the
set ω consisting of all the natural numbers. This means that we can already "count" farther than we could before; whereas before the only
numbers at our disposal were the elements of ω, now we have ω itself. We
have also the successor ω^+ of ω; this set is ordered in the obvious way, and,
moreover, the obvious ordering is a well ordering that satisfies the condition imposed on ordinal numbers. Indeed, if $\xi \in \omega^+$, then, by the definition of successor, either $\xi \in \omega$, in which case we already know that $s(\xi) = \xi$,
or else $\xi = \omega$, in which case $s(\xi) = \omega$, by the definition of order, so that
again $s(\xi) = \xi$. The argument just presented is quite general; it proves
that if α is an ordinal number, then so is α^+. It follows that our counting
process extends now up to and including ω, and ω^+, and $(\omega^+)^+$, and so on
ad infinitum.

At this point we make contact with our earlier discussion of what happens beyond ω. The axiom of substitution implies easily that there exists a unique function F on ω such that $F(0) = \omega$ and $F(n^+) = (F(n))^+$ for each natural number n. The range of this function is a set of interest for us; a set of even greater importance is the union of the set ω with the range of the function F. For reasons that will become clear only after we have at least glanced at the arithmetic of ordinal numbers, that union is usually denoted by $\omega 2$. If, borrowing again from the notation of ordinal arithmetic, we write $\omega + n$ for $F(n)$, then we can describe the set $\omega 2$ as the set consisting of all n (with n in ω) and of all $\omega + n$ (with n in ω).

It is now easy to verify that $\omega 2$ is an ordinal number. The verification depends, of course, on the definition of order in $\omega 2$. At this point both that definition and the proof are left as exercises; our official attention turns to some general remarks that include the facts about $\omega 2$ as easy special cases.

An order (partial or total) in a set X is uniquely determined by its initial segments. If, in other words, R and S are orders in X, and if, for each x in X, the set of all R-predecessors of x is the same as the set of all S-predecessors of x, then R and S are the same. This assertion is obvious whether predecessors are taken in the strict sense or not. The assertion applies, in particular, to well ordered sets. From this special case we infer that if it is possible at all to well order a set so as to make it an ordinal number, then there is only one way to do so. The set alone tells us what the relation that makes it an ordinal number must be; if that relation satisfies the requirements, then the set is an ordinal number, and otherwise it is not. To say that $s(\xi) = \xi$ means that the predecessors of ξ must be just the elements of ξ. The relation in question is therefore simply the relation of belonging. If $\eta < \xi$ is defined to mean $\eta \, \epsilon \, \xi$ whenever ξ and η are elements of a set α, then the result either is or is not a well ordering of α such that $s(\xi) = \xi$ for each ξ in α, and α is an ordinal number in the one case and not in the other.

We conclude this preliminary discussion of ordinal numbers by mentioning the names of the first few of them. After $0, 1, 2, \cdots$ comes ω, and after $\omega, \omega + 1, \omega + 2, \cdots$ comes $\omega 2$. After $\omega 2 + 1$ (that is, the successor of $\omega 2$) comes $\omega 2 + 2$, and then $\omega 2 + 3$; next after all the terms of the sequence so begun comes $\omega 3$. (Another application of the axiom of substitution is needed at this point.) Next come $\omega 3 + 1, \omega 3 + 2, \omega 3 + 3, \cdots$, and after them comes $\omega 4$. In this way we get successively $\omega, \omega 2, \omega 3, \omega 4, \cdots$. An application of the axiom of substitution yields something that follows them all in the same sense in which ω follows the natural numbers; that something is ω^2. After that the whole thing starts over again: $\omega^2 + 1$, $\omega^2 + 2$, $\cdots$, $\omega^2 + \omega$, $\omega^2 + \omega + 1$, $\omega^2 + \omega + 2$, $\cdots$, $\omega^2 + \omega 2$, $\omega^2 + \omega 2 + 1$, $\cdots$, $\omega^2 + \omega 3$, $\cdots$, $\omega^2 + \omega 4$, $\cdots$, $\omega^2 2$, $\cdots$, $\omega^2 3$, $\cdots$, ω^3, $\cdots$, ω^4, $\cdots$, ω^ω, $\cdots$, $\omega^{(\omega^\omega)}$, $\cdots$, $\omega^{(\omega^{(\omega^\omega)})}$, $\cdots$. The next one after all that is ε_0; then come $\varepsilon_0 + 1$, $\varepsilon_0 + 2$, $\cdots$, $\varepsilon_0 + \omega$, $\cdots$, $\varepsilon_0 + \omega 2$, $\cdots$, $\varepsilon_0 + \omega^2$, $\cdots$, $\varepsilon_0 + \omega^\omega$, $\cdots$, $\varepsilon_0 2$, $\cdots$, $\varepsilon_0 \omega$, $\cdots$, $\varepsilon_0 \omega^\omega$, $\cdots$, ε_0^2, $\cdots \cdots \cdots \cdots$.

SECTION 20

SETS OF ORDINAL NUMBERS

An ordinal number is, by definition, a special kind of well ordered set; we proceed to examine its special properties.

The most elementary fact is that each element of an ordinal number α is at the same time a subset of α. (In other words, every ordinal number is a transitive set.) Indeed, if $\xi \epsilon \alpha$, then the fact that $s(\xi) = \xi$ implies that each element of ξ is a predecessor of ξ in α and hence, in particular, an element of α.

If ξ is an element of an ordinal number α, then, as we have just seen, ξ is a subset of α, and, consequently, ξ is a well ordered set (with respect to the ordering it inherits from α). Assertion: ξ is in fact an ordinal number. Indeed, if $\eta \epsilon \xi$, then the initial segment determined by η in ξ is the same as the initial segment determined by η in α; since the latter is equal to η, so is the former. Another way of formulating the same result is to say that every initial segment of an ordinal number is an ordinal number.

The next thing to note is that if two ordinal numbers are similar, then they are equal. To prove this, suppose that α and β are ordinal numbers and that f is a similarity from α onto β; we shall show that $f(\xi) = \xi$ for each ξ in α. The proof is a straightforward transfinite induction. Write $S = \{\xi \epsilon \alpha : f(\xi) = \xi\}$. For each ξ in α, the least element of α that does not belong to $s(\xi)$ is ξ itself. Since f is a similarity, it follows that the least element of β that does not belong to the image of $s(\xi)$ under f is $f(\xi)$. These assertions imply that if $s(\xi) \subset S$, then $f(\xi)$ and ξ are ordinal numbers with the same initial segments, and hence that $f(\xi) = \xi$. We have proved thus that $\xi \epsilon S$ whenever $s(\xi) \subset S$. The principle of transfinite induction implies that $S = \alpha$, and from this it follows that $\alpha = \beta$.

If α and β are ordinal numbers, then, in particular, they are well ordered sets, and, consequently, either they are similar or else one of them is similar to an initial segment of the other. If, say, β is similar to an initial segment of α, then β is similar to an element of α. Since every element of α is an ordinal number, it follows that β *is* an element of α, or, in still other words, that α is a continuation of β. We know by now that if α and β are distinct ordinal numbers, then the statements

$$\beta \in \alpha,$$

$$\beta \subset \alpha,$$

α *is a continuation of* β,

are all equivalent to one another; if they hold, we may write

$$\beta < \alpha.$$

What we have just proved is that any two ordinal numbers are comparable; that is, if α and β are ordinal numbers, then either $\beta = \alpha$, or $\beta < \alpha$, or $\alpha < \beta$.

The result of the preceding paragraph can be expressed by saying that every set of ordinal numbers is totally ordered. In fact more is true: every set of ordinal numbers is well ordered. Suppose indeed that E is a nonempty set of ordinal numbers, and let α be an element of E. If $\alpha \leqq \beta$ for all β in E, then α is the first element of E and all is well. If this is not the case, then there exists an element β in E such that $\beta < \alpha$, i.e., $\beta \in \alpha$; in other words, then $\alpha \cap E$ is not empty. Since α is a well ordered set, $\alpha \cap E$ has a first element, say α_0. If $\beta \in E$, then either $\alpha \leqq \beta$ (in which case $\alpha_0 < \beta$), or $\beta < \alpha$ (in which case $\beta \in \alpha \cap E$ and therefore $\alpha_0 \leqq \beta$), and this proves that E has a first element, namely α_0.

Some ordinal numbers are finite; they are just the natural numbers (i.e., the elements of ω). The others are called *transfinite*; the set ω of all natural numbers is the smallest transfinite ordinal number. Each finite ordinal number (other than 0) has an immediate predecessor. If a transfinite ordinal number α has an immediate predecessor β, then, just as for natural numbers, $\alpha = \beta^+$. Not every transfinite ordinal number does have an immediate predecessor; the ones that do not are called *limit numbers*.

Suppose now that $\mathcal{C}$ is a collection of ordinal numbers. Since, as we have just seen, $\mathcal{C}$ is a continuation chain, it follows that the union α of the sets of $\mathcal{C}$ is a well ordered set such that for every ξ in $\mathcal{C}$, distinct from α itself, α is a continuation of ξ. The initial segment determined by an element in α is the same as the initial segment determined by that element whatever set of $\mathcal{C}$ it occurs in; this implies that α is an ordinal number. If $\xi \in \mathcal{C}$, then $\xi \leqq \alpha$; the number α is an upper bound of the elements of

e. If β is another upper bound of e, then $\xi \subset \beta$ whenever $\xi \in e$, and therefore, by the definition of unions, $\alpha \subset \beta$. This implies that α is the least upper bound of e; we have proved thus that every set of ordinal numbers has a supremum.

Is there a set that consists exactly of all the ordinal numbers? It is easy to see that the answer must be no. If there were such a set, then we could form the supremum of all ordinal numbers. That supremum would be an ordinal number greater than or equal to every ordinal number. Since, however, for each ordinal number there exists a strictly greater one (for example, its successor), this is impossible; it makes no sense to speak of the "set" of all ordinals. The contradiction, based on the assumption that there is such a set, is called the *Burali-Forti paradox*. (Burali-Forti was one man, not two.)

Our next purpose is to show that the concept of an ordinal number is not so special as it might appear, and that, in fact, each well ordered set resembles some ordinal number in all essential respects. "Resemblance" here is meant in the technical sense of similarity. An informal statement of the result is that each well ordered set can be counted.

Counting theorem. *Each well ordered set is similar to a unique ordinal number.*

PROOF. Since for ordinal numbers similarity is the same as equality, uniqueness is obvious. Suppose now that X is a well ordered set and suppose that an element a of X is such that the initial segment determined by each predecessor of a is similar to some (necessarily unique) ordinal number. If $S(x, \alpha)$ is the sentence that says "α is an ordinal number and $s(x) \cong \alpha$," then, for each x in $s(a)$, the set $\{\alpha : S(x, \alpha)\}$ can be formed; in fact, that set is a singleton. The axiom of substitution implies the existence of a set consisting exactly of the ordinal numbers similar to the initial segments determined by the predecessors of a. It follows, whether a is the immediate successor of one of its predecessors or the supremum of them all, that $s(a)$ is similar to an ordinal number. This argument prepares the way for an application of the principle of transfinite induction; the conclusion is that each initial segment in X is similar to some ordinal number. This fact, in turn, justifies another application of the axiom of substitution, just like the one made above; the final conclusion is, as desired, that X is similar to some ordinal number.

SECTION 21

ORDINAL ARITHMETIC

For natural numbers we used the recursion theorem to define the arithmetic operations, and, subsequently, we proved that those operations are related to the operations of set theory in various desirable ways. Thus, for instance, we know that the number of elements in the union of two disjoint finite sets E and F is equal to $\#(E) + \#(F)$. We observe now that this fact could have been used to define addition. If m and n are natural numbers, we could have defined their sum by finding disjoint sets E and F, with $\#(E) = m$ and $\#(F) = n$, and writing $m + n = \#(E \cup F)$.

Corresponding to what was done and to what could have been done for natural numbers, there are two standard approaches to ordinal arithmetic. Partly for the sake of variety, and partly because in this context recursion seems less natural, we shall emphasize the set-theoretic approach instead of the recursive one.

We begin by pointing out that there is a more or less obvious way of putting two well ordered sets together to form a new well ordered set. Informally speaking, the idea is to write down one of them and then to follow it by the other. If we try to say this rigorously, we immediately encounter the difficulty that the two sets may not be disjoint. When are we supposed to write down an element that is common to the two sets? The way out of the difficulty is to make the sets disjoint. This can be done by painting their elements different colors. In more mathematical language, replace the elements of the sets by those same elements taken together with some distinguishing object, using two different objects for the two sets. In completely mathematical language: if E and F are arbitrary sets, let $\hat{E}$ be the set of all ordered pairs $(x, 0)$ with x in E, and let $\hat{F}$ be the set of all ordered pairs $(x, 1)$ with x in F. The sets $\hat{E}$ and $\hat{F}$ are clearly disjoint. There is an obvious one-to-one correspondence between E and $\hat{E}$ $(x \to (x, 0))$ and another one between F and $\hat{F}$ $(x \to (x, 1))$.

These correspondences can be used to carry over whatever structure E and F may possess (for example, order) to $\hat{E}$ and $\hat{F}$. It follows that any time we are given two sets, with or without some additional structure, we may always replace them by disjoint sets with the same structure, and hence we may assume, with no loss of generality, that they were disjoint in the first place.

Before applying this construction to ordinal arithmetic, we observe that it can be generalized to arbitrary families of sets. If, indeed, $\{E_i\}$ is a family, write $\hat{E}_i$ for the set of all ordered pairs (x, i), with x in E_i. (In other words, $\hat{E}_i = E_i \times \{i\}$.) The family $\{\hat{E}_i\}$ is pairwise disjoint, and it can do anything the original family $\{E_i\}$ could do.

Suppose now that E and F are disjoint well ordered sets. Define order in $E \cup F$ so that pairs of elements in E, and also pairs of elements in F, retain the order they had, and so that each element of E precedes each element of F. (In ultraformal language: if R and S are the given order relations in E and F respectively, let $E \cup F$ be ordered by $R \cup S \cup (E \times F)$.) The fact that E and F were well ordered implies that $E \cup F$ is well ordered. The well ordered set $E \cup F$ is called the *ordinal sum* of the well ordered sets E and F.

There is an easy and worth while way of extending the concept of ordinal sum to infinitely many summands. Suppose that $\{E_i\}$ is a disjoint family of well ordered sets indexed by a well ordered set I. The ordinal sum of the family is the union $\bigcup_i E_i$, ordered as follows. If a and b are elements of the union, with $a \in E_i$ and $b \in E_j$, then $a < b$ means that either $i < j$ or else $i = j$ and a precedes b in the given order of E_i.

The definition of addition for ordinal numbers is now child's play. For each well ordered set X, let ord X be the unique ordinal number similar to X. (If X is finite, then ord X is the same as the natural number $\#(X)$ defined earlier.) If α and β are ordinal numbers, let A and B be disjoint well ordered sets with ord $A = \alpha$ and ord $B = \beta$, and let C be the ordinal sum of A and B. The *sum* $\alpha + \beta$ is, by definition, the ordinal number of C, so that ord A + ord B = ord C. It is important to note that the sum $\alpha + \beta$ is independent of the particular choice of the sets A and B; any other pair of disjoint sets, with the same ordinal numbers, would have given the same result.

These considerations extend without difficulty to the infinite case. If $\{\alpha_i\}$ is a well ordered family of ordinal numbers indexed by a well ordered set I, let $\{A_i\}$ be a disjoint family of well ordered sets with ord $A_i = \alpha_i$ for each i, and let A be the ordinal sum of the family $\{A_i\}$. The sum $\sum_{i \in I}$ ord A_i is, by definition, the ordinal number of A, so that $\sum_{i \in I}$ ord A_i

= ord A. Here too the final result is independent of the arbitrary choice of the well ordered sets A_i; any other choices (with the same ordinal numbers) would have given the same sum.

Some of the properties of addition for ordinal numbers are good and others are bad. On the good side of the ledger are the identities

$$\alpha + 0 = \alpha,$$
$$0 + \alpha = \alpha,$$
$$\alpha + 1 = \alpha^+,$$

and the associative law

$$\alpha + (\beta + \gamma) = (\alpha + \beta) + \gamma.$$

Equally laudable is the fact that $\alpha < \beta$ if and only if there exists an ordinal number γ different from 0 such that $\beta = \alpha + \gamma$. The proofs of all these assertions are elementary.

Almost all the bad behavior of addition stems from the failure of the commutative law. Sample: $1 + \omega = \omega$ (but, as we saw just above, $\omega + 1 \neq \omega$). The misbehavior of addition expresses some intuitively clear facts about order. If, for instance, we tack a new element in front of an infinite sequence (of type ω), the result is clearly similar to what we started with, but if we tack it on at the end instead, then we have ruined similarity; the old set had no last element but the new set has one.

The main use of infinite sums is to motivate and facilitate the study of products. If A and B are well ordered sets, it is natural to define their product as the result of adding A to itself B times. To make sense out of this, we must first of all manufacture a disjoint family of well ordered sets, each of which is similar to A, indexed by the set B. The general prescription for doing this works well here; all we need to do is to write $A_b = A \times \{b\}$ for each b in B. If now we examine the definition of ordinal sum as it applies to the family $\{A_b\}$, we are led to formulate the following definition. The *ordinal product* of two well ordered sets A and B is the Cartesian product $A \times B$ with the reverse lexicographic order. In other words, if (a, b) and (c, d) are in $A \times B$, then $(a, b) < (c, d)$ means that either $b < d$ or else $b = d$ and $a < c$.

If α and β are ordinal numbers, let A and B be well ordered sets with ord $A = \alpha$ and ord $B = \beta$, and let C be the ordinal product of A and B. The *product* $\alpha\beta$ is, by definition, the ordinal number of C, so that (ord A)(ord B) = ord C. The product is unambiguously defined, independently of the arbitrary choice of the well ordered sets A and B. Alternatively, at this point we could have avoided any arbitrariness at all by recalling that the most easily available well ordered set whose ordinal number is α is the ordinal number α itself (and similarly for β).

Like addition, multiplication has its good and bad properties. Among the good ones are the identities

$$\alpha 0 = 0,$$
$$0\alpha = 0,$$
$$\alpha 1 = \alpha,$$
$$1\alpha = \alpha,$$

the associative law

$$\alpha(\beta\gamma) = (\alpha\beta)\gamma,$$

the left distributive law

$$\alpha(\beta + \gamma) = \alpha\beta + \alpha\gamma,$$

and the fact that if the product of two ordinal numbers is zero, then one of the factors must be zero. (Note that we use the standard convention about multiplication taking precedence over addition; $\alpha\beta + \alpha\gamma$ denotes $(\alpha\beta) + (\alpha\gamma)$.)

The commutative law for multiplication fails, and so do many of its consequences. Thus, for instance, $2\omega = \omega$ (think of an infinite sequence of ordered pairs), but $\omega 2 \neq \omega$ (think of an ordered pair of infinite sequences). The right distributive law also fails; that is $(\alpha + \beta)\gamma$ is in general different from $\alpha\gamma + \beta\gamma$. Example: $(1 + 1)\omega = 2\omega = \omega$, but $1\omega + 1\omega = \omega + \omega = \omega 2$.

Just as repeated addition led to the definition of ordinal products, repeated multiplication could be used to define ordinal exponents. Alternatively, exponentiation can be approached via transfinite recursion. The precise details are part of an extensive and highly specialized theory of ordinal numbers. At this point we shall be content with hinting at the definition and mentioning its easiest consequences. To define α^β (where α and β are ordinal numbers), use definition by transfinite induction (on β). Begin by writing $\alpha^0 = 1$ and $\alpha^{\beta+1} = \alpha^\beta\alpha$; if β is a limit number, define α^β as the supremum of the numbers of the form α^γ, where $\gamma < \beta$. If this sketch of a definition is formulated with care, it follows that

$$0^\alpha = 0 \ (\alpha \geqq 1),$$
$$1^\gamma = 1,$$
$$\alpha^{\beta+\gamma} = \alpha^\beta\alpha^\gamma,$$
$$\alpha^{\beta\gamma} = (\alpha^\beta)^\gamma.$$

Not all the familiar laws of exponents hold; thus, for instance, $(\alpha\beta)^\gamma$ is in general different from $\alpha^\gamma\beta^\gamma$. Example: $(2\cdot2)^\omega = 4^\omega = \omega$, but $2^\omega \cdot 2^\omega = \omega \cdot \omega = \omega^2$.

Warning: the exponent notation for ordinal numbers, here and below, is not consistent with our earlier use of it. The unordered set 2^ω of all functions from ω to 2, and the well ordered set 2^ω that is the least upper bound of the sequence of ordinal numbers 2, $2\cdot2$, $2\cdot2\cdot2$, etc., are not the same thing at all. There is no help for it; mathematical usage is firmly established in both camps. If, in a particular situation, the context does not reveal which of the two interpretations is to be used, then explicit verbal indication must be given.

SECTION 22

THE SCHRÖDER–BERNSTEIN THEOREM

The purpose of counting is to compare the size of one set with that of another; the most familiar method of counting the elements of a set is to arrange them in some appropriate order. The theory of ordinal numbers is an ingenious abstraction of the method, but it falls somewhat short of achieving the purpose. This is not to say that ordinal numbers are useless; it just turns out that their main use is elsewhere, in topology, for instance, as a source of illuminating examples and counterexamples. In what follows we shall continue to pay some attention to ordinal numbers, but they will cease to occupy the center of the stage. (It is of some importance to know that we could in fact dispense with them altogether. The theory of cardinal numbers can be constructed with the aid of ordinal numbers, or without it; both kinds of constructions have advantages.) With these prefatory remarks out of the way, we turn to the problem of comparing the sizes of sets.

The problem is to compare the sizes of sets when their elements do not appear to have anything to do with each other. It is easy enough to decide that there are more people in France than in Paris. It is not quite so easy, however, to compare the age of the universe in seconds with the population of Paris in electrons. For some mathematical examples, consider the following pairs of sets, defined in terms of an auxiliary set A: (i) $X = A$, $Y = A^+$; (ii) $X = \mathcal{O}(A)$, $Y = 2^A$; (iii) X is the set of all one-to-one mappings of A into itself, Y is the set of all finite subsets of A. In each case we may ask which of the two sets X and Y has more elements. The problem is first to find a rigorous interpretation of the question and then to answer it.

The well ordering theorem tells us that every set can be well ordered. For well ordered sets we have what seems to be a reasonable measure of size, namely, their ordinal number. Do these two remarks solve the problem? To compare the sizes of X and Y, may we just well order each of them and then compare ord X and ord Y? The answer is most emphatically no. The trouble is that one and the same set can be well ordered in many ways. The ordinal number of a well ordered set measures the well ordering more than it measures the set. For a concrete example consider the set ω of all natural numbers. Introduce a new order by placing 0 after everything else. (In other words, if n and m are non-zero natural numbers, then arrange them in their usual order; if, however, $n = 0$ and $m \neq 0$, let m precede n.) The result is a well ordering of ω; the ordinal number of this well ordering is $\omega + 1$.

If X and Y are well ordered sets, then a necessary and sufficient condition that ord $X <$ ord Y is that X be similar to an initial segment of Y. It follows that we could compare the ordinal sizes of two well ordered sets even without knowing anything about ordinal numbers; all we would need to know is the concept of similarity. Similarity was defined for ordered sets; the central concept for arbitrary unordered sets is that of equivalence. (Recall that two sets X and Y are called equivalent, $X \sim Y$, in case there exists a one-to-one correspondence between them.) If we replace similarity by equivalence, then something like the suggestion of the preceding paragraph becomes usable. The point is that we do not have to know what size is if all we want is to compare sizes.

If X and Y are sets such that X is equivalent to a subset of Y, we shall write

$$X \precsim Y.$$

The notation is temporary and does not deserve a permanent name. As long as it lasts, however, it is convenient to have a way of referring to it; a reasonable possibility is to say that Y *dominates* X. The set of those ordered pairs (X, Y) of subsets of some set E for which $X \precsim Y$ constitutes a relation in the power set of E. The symbolism correctly suggests some of the properties of the concept that it denotes. Since the symbolism is reminiscent of partial orders, and since a partial order is reflexive, antisymmetric, and transitive, we may expect that domination has similar properties.

Reflexivity and transitivity cause no trouble. Since each set X is equivalent to a subset (namely, X) of itself, it follows that $X \precsim X$ for all X. If f is a one-to-one correspondence between X and a subset of Y, and if g is a one-to-one correspondence between Y and a subset of Z, then we may restrict g to the range of f and compound the result with f; the conclusion is that X is equivalent to a subset of Z. In other words, if $X \precsim Y$ and $Y \precsim Z$, then $X \precsim Z$.

The interesting question is that of antisymmetry. If $X \precsim Y$ and $Y \precsim X$, can we conclude that $X = Y$? This is absurd; the assumptions are satisfied whenever X and Y are equivalent, and equivalent sets need not be identical. What then can we say about two sets if all we know is that each of them is equivalent to a subset of the other? The answer is contained in the following celebrated and important result.

Schröder-Bernstein theorem. *If $X \precsim Y$ and $Y \precsim X$, then $X \sim Y$.*

REMARK. Observe that the converse, which is incidentally a considerable strengthening of the assertion of reflexivity, follows trivially from the definition of domination.

PROOF. Let f be a one-to-one mapping from X into Y and let g be a one-to-one mapping from Y into X; the problem is to construct a one-to-one correspondence between X and Y. It is convenient to assume that the sets X and Y have no elements in common; if that is not true, we can so easily make it true that the added assumption involves no loss of generality.

We shall say that an element x in X is the *parent* of the element $f(x)$ in Y, and, similarly, that an element y in Y is the parent of $g(y)$ in X. Each element x of X has an infinite sequence of *descendants*, namely, $f(x)$, $g(f(x))$, $f(g(f(x)))$, etc., and similarly, the descendants of an element y of Y are $g(y)$, $f(g(y))$, $g(f(g(y)))$, etc. This definition implies that each term in the sequence is a descendant of all preceding terms; we shall also say that each term in the sequence is an *ancestor* of all following terms.

For each element (in either X or Y) one of three things must happen. If we keep tracing the ancestry of the element back as far as possible, then either we ultimately come to an element of X that has no parent (these orphans are exactly the elements of $X - g(Y)$), or we ultimately come to an element of Y that has no parent ($Y - f(X)$), or the lineage regresses ad infinitum. Let X_X be the set of those elements of X that originate in X (i.e., X_X consists of the elements of $X - g(Y)$ together with all their descendants in X), let X_Y be the set of those elements of X that originate in Y (i.e., X_Y consists of all the descendants in X of the elements of $Y - f(X)$), and let X_∞ be the set of those elements of X that have no parentless ancestor. Partition Y similarly into the three sets Y_X, Y_Y, and Y_∞.

If $x \in X_X$, then $f(x) \in Y_X$, and, in fact, the restriction of f to X_X is a one-to-one correspondence between X_X and Y_X. If $x \in X_Y$, then x belongs to the domain of the inverse function g^{-1} and $g^{-1}(x) \in Y_Y$; in fact the restriction of g^{-1} to X_Y is a one-to-one correspondence between X_Y and Y_Y. If, finally, $x \in X_\infty$, then $f(x) \in Y_\infty$, and the restriction of f to X_∞ is a one-to-one correspondence between X_∞ and Y_∞; alternatively, if $x \in X_\infty$, then $g^{-1}(x) \in Y_\infty$, and the restriction of g^{-1} to X_∞ is a one-to-one correspondence between X_∞ and Y_∞. By combining these three one-to-one correspondences, we obtain a one-to-one correspondence between X and Y.

EXERCISE. Suppose that f is a mapping from X into Y and g is a mapping from Y into X. Prove that there exist subsets A and B of X and Y respectively, such that $f(A) = B$ and $g(Y - B) = X - A$. This result can be used to give a proof of the Schröder-Bernstein theorem that looks quite different from the one above.

By now we know that domination has the essential properties of a partial order; we conclude this introductory discussion by observing that the order is in fact total. The assertion is known as the comparability theorem for sets: it says that if X and Y are sets, then either $X \precsim Y$ or $Y \precsim X$. The proof is an immediate consequence of the well ordering theorem and of the comparability theorem for well ordered sets. Well order both X and Y and use the fact that either the well ordered sets so obtained are similar or one of them is similar to an initial segment of the other; in the former case X and Y are equivalent, and in the latter one of them is equivalent to a subset of the other.

SECTION 23

COUNTABLE SETS

If X and Y are sets such that Y dominates X and X dominates Y, then the Schröder-Bernstein theorem applies and says that X is equivalent to Y. If Y dominates X but X does not dominate Y, so that X is not equivalent to Y, we shall write

$$X \prec Y,$$

and we shall say that Y *strictly dominates* X.

Domination and strict domination can be used to express some of the facts about finite and infinite sets in a neat form. Recall that a set X is called finite in case it is equivalent to some natural number; otherwise it is infinite. We know that if $X \precsim Y$ and Y is finite, then X is finite, and we know that ω is infinite (§ 13); we know also that if X is infinite, then $\omega \precsim X$ (§ 15). The converse of the last assertion is true and can be proved either directly (using the fact that a finite set cannot be equivalent to a proper subset of itself) or as an application of the Schröder-Bernstein theorem. (If $\omega \precsim X$, then it is impossible that there exist a natural number n such that $X \sim n$, for then we should have $\omega \precsim n$, and that contradicts the fact that ω is infinite.)

We have just seen that a set X is infinite if and only if $\omega \precsim X$; next we shall prove that X is finite if and only if $X \prec \omega$. The proof depends on the transitivity of strict domination: if $X \precsim Y$ and $Y \precsim Z$, and if at least one of these dominations is strict, then $X \prec Z$. Indeed, clearly, $X \precsim Z$. If we had $Z \precsim X$, then we should have $Y \precsim X$ and $Z \precsim Y$ and hence (by the Schröder-Bernstein theorem) $X \sim Y$ and $Y \sim Z$, in contradiction to the assumption of strict domination. If now X is finite, then $X \sim n$ for some natural number n, and, since ω is infinite, $n \prec \omega$, so that $X \prec \omega$.

If, conversely, $X \prec \omega$, then X must be finite, for otherwise we should have $\omega \precsim X$, and hence $\omega \prec \omega$, which is absurd.

A set X is called *countable* (or *denumerable*) in case $X \precsim \omega$ and *countably infinite* in case $X \sim \omega$. Clearly a countable set is either finite or countably infinite. Our main purpose in the immediate sequel is to show that many set-theoretic constructions when performed on countable sets lead again to countable sets.

We begin with the observation that every subset of ω is countable, and we go on to deduce that every subset of each countable set is countable. These facts are trivial but useful.

If f is a function from ω *onto* a set X, then X is countable. For the proof, observe that for each x in X the set $f^{-1}(\{x\})$ is not empty (this is where the *onto* character of f is important), and consequently, for each x in X, we may find a natural number $g(x)$ such that $f(g(x)) = x$. Since the function g is a one-to-one mapping from X into ω, this proves that $X \precsim \omega$. The reader who worries about such things might have noticed that this proof made use of the axiom of choice, and he may want to know that there is an alternative proof that does not depend on that axiom. (There is.) The same comment applies on a few other occasions in this section and its successors but we shall refrain from making it.

It follows from the preceding paragraph that a set X is countable if and only if there exists a function from some countable set onto X. A closely related result is this: if Y is any particular countably infinite set, then a necessary and sufficient condition that a non-empty set X be countable is that there exist a function from Y onto X.

The mapping $n \rightarrow 2n$ is a one-to-one correspondence between ω and the set A of all even numbers, so that A is countably infinite. This implies that if X is a countable set, then there exists a function f that maps A onto X. Since, similarly, the mapping $n \rightarrow 2n + 1$ is a one-to-one correspondence between ω and the set B of all odd numbers, it follows that if Y is a countable set, then there exists a function g that maps B onto Y. The function h that agrees with f on A and with g on B (i.e., $h(x) = f(x)$ when $x \in A$ and $h(x) = g(x)$ when $x \in B$) maps ω onto $X \cup Y$. Conclusion: the union of two countable sets is countable. From here on an easy argument by mathematical induction proves that the union of a finite set of countable sets is countable. The same result can be obtained by imitating the trick that worked for two sets; the basis of the method is the fact that for each non-zero natural number n there exists a pairwise disjoint family $\{A_i\}$ ($i < n$) of infinite subsets of ω whose union is equal to ω.

The same method can be used to prove still more. Assertion: there exists a pairwise disjoint family $\{A_n\}$ ($n \in \omega$) of infinite subsets of ω whose union is equal to ω. One way to prove this is to write down the elements of ω in an infinite array by counting down the diagonals, thus:

0	1	3	6	10	15	···
2	4	7	11	16	···	
5	8	12	17	···		
9	13	18	···			
14	19	···				
20	···					
···						

and then to consider the sequence of the rows of this array. Another way is to let A_0 consist of 0 and the odd numbers, let A_1 be the set obtained by doubling each non-zero element of A_0, and, inductively, let A_{n+1} be the set obtained by doubling each element of A_n, $n \geqq 1$. Either way (and there are many others still) the details are easy to fill in. Conclusion: the union of a countably infinite family of countable sets is countable. Proof: given the family $\{X_n\}$ ($n \in \omega$) of countable sets, find a family $\{f_n\}$ of functions such that, for each n, the function f_n maps A_n onto X_n, and define a function f from ω onto $\bigcup_n X_n$ by writing $f(k) = f_n(k)$ whenever $k \in A_n$. This result combined with the result of the preceding paragraph implies that the union of a countable set of countable sets is always countable.

An interesting and useful corollary is that the Cartesian product of two countable sets is also countable. Since

$$X \times Y = \bigcup_{y \in Y} (X \times \{y\}),$$

and since, if X is countable, then, for each fixed y in Y, the set $X \times \{y\}$ is obviously countable (use the one-to-one correspondence $x \to (x, y)$), the result follows from the preceding paragraph.

EXERCISE. Prove that the set of all finite subsets of a countable set is countable. Prove that if every countable subset of a totally ordered set X is well ordered, then X itself is well ordered.

On the basis of the preceding discussion it would not be unreasonable to guess that every set is countable. We proceed to show that that is not so; this negative result is what makes the theory of cardinal numbers interesting.

Cantor's theorem. *Every set is strictly dominated by its power set, or, in other words,*

$$X < \mathcal{P}(X)$$

for all X.

PROOF. There is a natural one-to-one mapping from X into $\mathcal{P}(X)$, namely, the mapping that associates with each element x of X the singleton $\{x\}$. The existence of this mapping proves that $X \lesssim \mathcal{P}(X)$; it remains to prove that X is not equivalent to $\mathcal{P}(X)$.

Assume that f is a one-to-one mapping from X onto $\mathcal{P}(X)$; our purpose is to show that this assumption leads to a contradiction. Write $A = \{x \in X : x \in' f(x)\}$; in words, A consists of those elements of X that are not contained in the corresponding set. Since $A \in \mathcal{P}(X)$ and since f maps X onto $\mathcal{P}(X)$, there exists an element a in X such that $f(a) = A$. The element a either belongs to the set A or it does not. If $a \in A$, then, by the definition of A, we must have $a \in' f(a)$, and since $f(a) = A$ this is impossible. If $a \in' A$, then, again by the definition of A, we must have $a \in f(a)$, and this too is impossible. The contradiction has arrived and the proof of Cantor's theorem is complete.

Since $\mathcal{P}(X)$ is always equivalent to 2^X (where 2^X is the set of all functions from X into 2), Cantor's theorem implies that $X < 2^X$ for all X. If in particular we take ω in the role of X, then we may conclude that the set of all sets of natural numbers is *uncountable* (i.e., not countable, non-denumerable), or, equivalently, that 2^ω is uncountable. Here 2^ω is the set of all infinite sequences of 0's and 1's (i.e., functions from ω into 2). Note that if we interpret 2^ω in the sense of ordinal exponentiation, then 2^ω is countable (in fact $2^\omega = \omega$).

SECTION 24

CARDINAL ARITHMETIC

One result of our study of the comparative sizes of sets will be to define a new concept, called *cardinal number,* and to associate with each set X a cardinal number, denoted by card X. The definitions are such that for each cardinal number a there exist sets A with card $A = a$. We shall also define an ordering for cardinal numbers, denoted as usual by $\leq$. The connection between these new concepts and the ones already at our disposal is easy to describe: it will turn out that card $X = $ card Y if and only if $X \sim Y$, and card $X < $ card Y if and only if $X \prec Y$. (If a and b are cardinal numbers, $a < b$ means, of course, that $a \leq b$ but $a \neq b$.)

The definition of cardinal numbers can be approached in several different ways, each of which has its strong advocates. To keep the peace as long as possible, and to demonstrate that the essential properties of the concept are independent of the approach, we shall postpone the basic construction. We proceed, instead, to study the arithmetic of cardinal numbers. In the course of that study we shall make use of the connection, described above, between cardinal inequality and set domination; that much of a loan from the future will be enough for the purpose.

If a and b are cardinal numbers, and if A and B are disjoint sets with card $A = a$ and card $B = b$, we write, by definition, $a + b = $ card $(A \cup B)$. If C and D are disjoint sets with card $C = a$ and card $D = b$, then $A \sim C$ and $B \sim D$; it follows that $A \cup B \sim C \cup D$, and hence that $a + b$ is unambiguously defined, independently of the arbitrary choice of A and B. Cardinal addition, thus defined, is commutative $(a + b = b + a)$, and associative $(a + (b + c) = (a + b) + c)$; these identities are immediate consequences of the corresponding facts about the formation of unions.

EXERCISE. Prove that if a, b, c, and d are cardinal numbers such that $a \leqq b$ and $c \leqq d$, then $a + c \leqq b + d$.

There is no difficulty about defining addition for infinitely many summands. If $\{a_i\}$ is a family of cardinal numbers, and if $\{A_i\}$ is a correspondingly indexed family of pairwise disjoint sets such that card $A_i = a_i$ for each i, then we write, by definition,

$$\sum_i a_i = \text{card} \left(\bigcup_i A_i \right).$$

As before, the definition is unambiguous.

To define the product ab of two cardinal numbers a and b, we find sets A and B with card $A = a$ and card $B = b$, and we write $ab = \text{card} (A \times B)$. The replacement of A and B by equivalent sets yields the same value of the product. Alternatively, we could have defined ab by "adding a to itself b times"; this refers to the formation of the infinite sum $\sum_{i \epsilon I} a_i$, where the index set I has cardinal number b, and where $a_i = a$ for each i in I. The reader should have no difficulty in verifying that this proposed alternative definition is indeed equivalent to the one that uses Cartesian products. Cardinal multiplication is commutative ($ab = ba$) and associative ($a(bc) = (ab)c$), and multiplication distributes over addition ($a(b + c) = ab + ac$); the proofs are elementary.

EXERCISE. Prove that if a, b, c, and d are cardinal numbers such that $a \leqq b$ and $c \leqq d$, then $ac \leqq bd$.

There is no difficulty about defining multiplication for infinitely many factors. If $\{a_i\}$ is a family of cardinal numbers, and if $\{A_i\}$ is a correspondingly indexed family of sets such that card $A_i = a_i$ for each i, then we write, by definition,

$$\prod_i a_i = \text{card} \left(\times_i A_i \right).$$

The definition is unambiguous.

EXERCISE. If $\{a_i\}$ ($i \epsilon I$) and $\{b_i\}$ ($i \epsilon I$) are families of cardinal numbers such that $a_i < b_i$ for each i in I, then $\sum_i a_i < \prod_i b_i$.

We can go from products to exponents the same way as we went from sums to products. The definition of a^b, for cardinal numbers a and b, is most profitably given directly, but an alternative approach goes via repeated multiplication. For the direct definition, find sets A and B with card $A = a$ and card $B = b$, and write $a^b = \text{card} A^B$. Alternatively, to define a^b "multiply a by itself b times." More precisely: form $\prod_{i \epsilon I} a_i$, where the index set I has cardinal number b, and where $a_i = a$ for each i

in I. The familiar laws of exponents hold. That is, if a, b, and c are cardinal numbers, then

$$a^{b+c} = a^b a^c,$$

$$(ab)^c = a^c b^c,$$

$$a^{bc} = (a^b)^c.$$

EXERCISE. Prove that if a, b, and c are cardinal numbers such that $a \leq b$, then $a^c \leq b^c$. Prove that if a and b are finite, greater than 1, and if c is infinite, then $a^c = b^c$.

The preceding definitions and their consequences are reasonably straightforward and not at all surprising. If they are restricted to finite sets only, the result is the familiar finite arithmetic. The novelty of the subject arises in the formation of sums, products, and powers in which at least one term is infinite. The words "finite" and "infinite" are used here in a very natural sense: a cardinal number is *finite* if it is the cardinal number of a finite set, and *infinite* otherwise.

If a and b are cardinal numbers such that a is finite and b is infinite, then

$$a + b = b.$$

For the proof, suppose that A and B are disjoint sets such that A is equivalent to some natural number k and B is infinite; we are to prove that $A \cup B \sim B$. Since $\omega \lesssim B$, we may and do assume that $\omega \subset B$. We define a mapping f from $A \cup B$ to B as follows: the restriction of f to A is a one-to-one correspondence between A and k, the restriction of f to ω is given by $f(n) = n + k$ for all n, and the restriction of f to $B - \omega$ is the identity mapping on $B - \omega$. Since the result is a one-to-one correspondence between $A \cup B$ and B, the proof is complete.

Next: if a is an infinite cardinal number, then

$$a + a = a.$$

For the proof, let A be a set with card $A = a$. Since the set $A \times 2$ is the union of two disjoint sets equivalent to A (namely, $A \times \{0\}$ and $A \times \{1\}$), it would be sufficient to prove that $A \times 2$ is equivalent to A. The approach we shall use will not quite prove that much, but it will come close enough. The idea is to approximate the construction of the desired one-to-one correspondence by using larger and larger subsets of A.

Precisely speaking, let $\mathcal{F}$ be the collection of all functions f such that the domain of f is of the form $X \times 2$, for some subset X of A, and such that f is a one-to-one correspondence between $X \times 2$ and X. If X is a countably infinite subset of A, then $X \times 2 \sim X$. This implies that the collection $\mathfrak{F}$ is not empty; at the very least it contains the one-to-one correspondences between $X \times 2$ and X for the countably infinite subsets X of A. The collection $\mathfrak{F}$ is partially ordered by extension. Since a straightforward verification shows that the hypotheses of Zorn's lemma are satisfied, it follows that $\mathfrak{F}$ contains a maximal element f with ran $f = X$, say.

Assertion: $A - X$ is finite. If $A - X$ were infinite, then it would include a countably infinite set, say Y. By combining f with a one-to-one correspondence between $Y \times 2$ and Y we could obtain a proper extension of f, in contradiction to the assumed maximality.

Since card $X +$ card $X =$ card X, and since card $A =$ card $X +$ card $(A - X)$, the fact that $A - X$ is finite completes the proof that card $A +$ card $A =$ card A.

Here is one more result in additive cardinal arithmetic: if a and b are cardinal numbers at least one of which is infinite, and if c is equal to the larger one of a and b, then

$$a + b = c.$$

Suppose that b is infinite, and let A and B be disjoint sets with card $A = a$ and card $B = b$. Since $a \leq c$ and $b \leq c$, it follows that $a + b \leq c + c$, and since $c \leq$ card $(A \cup B)$, it follows that $c \leq a + b$. The result follows from the antisymmetry of the ordering of cardinal numbers.

The principal result in multiplicative cardinal arithmetic is that if a is an infinite cardinal number, then

$$a \cdot a = a.$$

The proof resembles the proof of the corresponding additive fact. Let $\mathfrak{F}$ be the collection of all functions f such that the domain of f is of the form $X \times X$ for some subset X of A, and such that f is a one-to-one correspondence between $X \times X$ and X. If X is a countably infinite subset of A, then $X \times X \sim X$. This implies that the collection $\mathfrak{F}$ is not empty; at the very least it contains the one-to-one correspondences between $X \times X$ and X for the countably infinite subsets X of A. The collection $\mathfrak{F}$ is partially ordered by extension. The hypotheses of Zorn's lemma are easily verified, and it follows that $\mathfrak{F}$ contains a maximal element f with ran $f = X$, say. Since (card X)(card X) $=$ card X, the proof may be completed by showing that card $X =$ card A.

Assume that card $X <$ card A. Since card A is equal to the larger one of card X and card $(A - X)$, this implies that card $A =$ card $(A - X)$, and hence that card $X <$ card $(A - X)$. From this it follows that $A - X$

has a subset Y equivalent to X. Since each of the disjoint sets $X \times Y$, $Y \times X$, and $Y \times Y$ is infinite and equivalent to $X \times X$, hence to X, and hence to Y, it follows that their union is equivalent to Y. By combining f with a one-to-one correspondence between that union and Y, we obtain a proper extension of f, in contradiction to the assumed maximality. This implies that our present hypothesis (card $X <$ card A) is untenable and hence completes the proof.

EXERCISE. Prove that if a and b are cardinal numbers at least one of which is infinite, then $a + b = ab$. Prove that if a and b are cardinal numbers such that a is infinite and b is finite, then $a^b = a$.

SECTION 25

CARDINAL NUMBERS

We know quite a bit about cardinal numbers by now, but we still do not know what they are. Speaking vaguely, we may say that the cardinal number of a set is the property that the set has in common with all sets equivalent to it. We may try to make this precise by saying that the cardinal number of X is equal to the set of all sets equivalent to X, but the attempt will fail; there is no set as large as that. The next thing to try, suggested by analogy with our approach to the definition of natural numbers, is to define the cardinal number of a set X as some particular carefully selected set equivalent to X. This is what we proceed to do.

For each set X there are too many other sets equivalent to X; our first problem is to narrow the field. Since we know that every set is equivalent to some ordinal number, it is not unnatural to look for the typical sets, the representative sets, among ordinal numbers.

To be sure, a set can be equivalent to many ordinal numbers. A hopeful sign, however, is the fact that, for each set X, the ordinal numbers equivalent to X constitute a set. To prove this, observe first that it is easy to produce an ordinal number that is surely greater, strictly greater, than all the ordinal numbers equivalent to X. Suppose in fact that γ is an ordinal number equivalent to the power set $\mathcal{P}(X)$. If α is an ordinal number equivalent to X, then the set α is strictly dominated by the set γ (i.e., card α < card γ). It follows that we cannot have $\gamma \leqq \alpha$, and, consequently, we must have $\alpha < \gamma$. Since, for ordinal numbers, $\alpha < \gamma$ means the same thing as $\alpha \in \gamma$, we have found a set, namely γ, that contains every ordinal number equivalent to X, and this implies that the ordinal numbers equivalent to X do constitute a set.

Which one among the ordinal numbers equivalent to X deserves to be singled out and called the cardinal number of X? The question has only one natural answer. Every set of ordinal numbers is well ordered; the

99

least element of a well ordered set is the only one that seems to clamor for special attention.

We are now prepared for the definition: a *cardinal number* is an ordinal number α such that if β is an ordinal number equivalent to α (i.e., card $\alpha = $ card β), then $\alpha \leq \beta$. The ordinal numbers with this property have also been called *initial numbers*. If X is a set, then card X, the cardinal number of X (also known as the *power* of X), is the least ordinal number equivalent to X.

EXERCISE. Prove that each infinite cardinal number is a limit number.

Since each set is equivalent to its cardinal number, it follows that if card $X = $ card Y, then $X \sim Y$. If, conversely, $X \sim Y$, then card $X \sim$ card Y. Since card X is the least ordinal number equivalent to X, it follows that card $X \leq $ card Y, and, since the situation is symmetric in X and Y, we also have card $Y \leq $ card X. In other words card $X = $ card Y if and only if $X \sim Y$; this was one of the conditions on cardinal numbers that we needed in the development of cardinal arithmetic.

A finite ordinal number (i.e., a natural number) is not equivalent to any finite ordinal number distinct from itself. It follows that if X is finite, then the set of ordinal numbers equivalent to X is a singleton, and, consequently, the cardinal number of X is the same as the ordinal number of X. Both cardinal numbers and ordinal numbers are generalizations of the natural numbers; in the familiar finite cases both the generalizations coincide with the special case that gave rise to them in the first place. As an almost trivial application of these remarks, we can now calculate the cardinal number of a power set $\mathcal{P}(A)$: if card $A = a$, then card $\mathcal{P}(A) = 2^a$. (Note that the result, though simple, could not have been stated before this; till now we did not know that 2 is a cardinal number.) The proof is immediate from the fact that $\mathcal{P}(A)$ is equivalent to 2^A.

If α and β are ordinal numbers, we know what it means to say that $\alpha < \beta$ or $\alpha \leq \beta$. It follows that cardinal numbers come to us automatically equipped with an order. The order satisfies the conditions we borrowed for our discussion of cardinal arithmetic. Indeed: if card $X < $ card Y, then card X is a subset of card Y, and it follows that $X \lesssim Y$. If we had $X \sim Y$, then, as we have already seen, we should have card $X = $ card Y; it follows that we must have $X < Y$. If, finally, $X < Y$, then it is impossible that card $Y \leq $ card X (for similarity implies equivalence), and hence card $X < $ card Y.

As an application of these considerations we mention the inequality

$$a < 2^a,$$

valid for all cardinal numbers a. Proof: if A is a set with card $A = a$, then $A \prec \mathcal{P}(A)$, hence card $A <$ card $\mathcal{P}(A)$, and therefore $a < 2^a$.

EXERCISE. If card $A = a$, what is the cardinal number of the set of all one-to-one mappings of A onto itself? What is the cardinal number of the set of all countably infinite subsets of A?

The facts about the ordering of ordinal numbers are at the same time facts about the ordering of cardinal numbers. Thus, for instance, we know that any two cardinal numbers are comparable (always either $a < b$, or $a = b$, or $b < a$), and that, in fact, every set of cardinal numbers is well ordered. We know also that every set of cardinal numbers has an upper bound (in fact, a supremum), and that, moreover, for every set of cardinal numbers, there is a cardinal number strictly greater than any of them. This implies of course that there is no largest cardinal number, or, equivalently, that there is no set that consists exactly of all the cardinal numbers. The contradiction, based on the assumption that there is such a set, is known as *Cantor's paradox*.

The fact that cardinal numbers are special ordinal numbers simplifies some aspects of the theory, but, at the same time, it introduces the possibility of some confusion that it is essential to avoid. One major source of difficulty is the notation for the arithmetic operations. If a and b are cardinal numbers, then they are also ordinal numbers, and, consequently, the sum $a + b$ has two possible meanings. The cardinal sum of two cardinal numbers is in general not the same as their ordinal sum. All this sounds worse than it is; in practice it is easy to avoid confusion. The context, the use of special symbols for cardinal numbers, and an occasional explicit warning can make the discussion flow quite smoothly.

EXERCISE. Prove that if α and β are ordinal numbers, then card $(\alpha + \beta)$ = card α + card β and card $(\alpha\beta)$ = (card α)(card β). Use the ordinal interpretation of the operations on the left side and the cardinal interpretation on the right.

One of the special symbols for cardinal numbers that is used very frequently is the first letter ($\aleph$, aleph) of the Hebrew alphabet. Thus in particular the smallest transfinite ordinal number, i.e., ω, is a cardinal number, and, as such, it is always denoted by $\aleph_0$.

Every one of the ordinal numbers that we have explicitly named so far is countable. In many of the applications of set theory an important role is played by the smallest uncountable ordinal number, frequently denoted by Ω. The most important property of ω is that it is an infinite well ordered set each of whose initial segments is finite; correspondingly, the most important property of Ω is that it is an uncountably infinite well ordered set each of whose initial segments is countable.

The least uncountable ordinal number Ω clearly satisfies the defining condition of a cardinal number; in its cardinal role it is always denoted by $\aleph_1$. Equivalently, $\aleph_1$ may be characterized as the least cardinal number strictly greater than $\aleph_0$, or, in other words, the immediate successor of $\aleph_0$ in the ordering of cardinal numbers.

The arithmetic relation between $\aleph_0$ and $\aleph_1$ is the subject of a famous old problem about cardinal numbers. How do we get from $\aleph_0$ to $\aleph_1$ by arithmetic operations? We know by now that the most elementary steps, involving sums and products, just lead from $\aleph_0$ back to $\aleph_0$ again. The simplest thing we know to do that starts with $\aleph_0$ and ends up with something larger is to form $2^{\aleph_0}$. We know therefore that $\aleph_1 \leqq 2^{\aleph_0}$. Is the inequality strict? Is there an uncountable cardinal number strictly less than $2^{\aleph_0}$? The celebrated *continuum hypothesis* asserts, as a guess, that the answer is no, or, in other words, that $\aleph_1 = 2^{\aleph_0}$. All that is known for sure is that the continuum hypothesis is consistent with the axioms of set theory.

For each infinite cardinal number a, consider the set $c(a)$ of all infinite cardinal numbers that are strictly less than a. If $a = \aleph_0$, then $c(a) = \varnothing$; if $a = \aleph_1$, then $c(a) = \{\aleph_0\}$. Since $c(a)$ is a well ordered set, it has an ordinal number, say α. The connection between a and α is usually expressed by writing $a = \aleph_\alpha$. An equivalent definition of the cardinal numbers $\aleph_\alpha$ proceeds by transfinite induction; according to that approach $\aleph_\alpha$ (for $\alpha > 0$) is the smallest cardinal number that is strictly greater than all the $\aleph_\beta$'s with $\beta < \alpha$. The *generalized continuum hypothesis* is the conjecture that $\aleph_{\alpha+1} = 2^{\aleph_\alpha}$ for each ordinal number α.

INDEX